LONDON MATHEMATICAL SOCIETY LECTURE NOTE SERIES

The books in the series listed below are available from booksellers, or, in case of difficulty, from Cambridge University Press.

17 Differential germs and catastrophes, Th. BROCKER & L. LANDER
34 Representation theory of Lie groups, M.F. ATIYAH *et al*
36 Homological group theory, C.T.C. WALL (ed)
39 Affine sets and affine groups, D.G. NORTHCOTT
40 Introduction to H_p spaces, P.J. KOOSIS
43 Graphs, codes and designs, P.J. CAMERON & J.H. VAN LINT
45 Recursion theory: its generalisations and applications, F.R. DRAKE & S.S. WAINER (eds)
46 p-adic analysis: a short course on recent work, N. KOBLITZ
49 Finite geometries and designs, P. CAMERON, J.W.P. HIRSCHFELD & D.R. HUGHES (eds)
50 Commutator calculus and groups of homotopy classes, H.J. BAUES
51 Synthetic differential geometry, A. KOCK
54 Markov processes and related problems of analysis, E.B. DYNKIN
57 Techniques of geometric topology, R.A. FENN
58 Singularities of smooth functions and maps, J.A. MARTINET
59 Applicable differential geometry, M. CRAMPIN & F.A.E. PIRANI
60 Integrable systems, S.P. NOVIKOV *et al*
62 Economics for mathematicians, J.W.S. CASSELS
65 Several complex variables and complex manifolds I, M.J. FIELD
66 Several complex variables and complex manifolds II, M.J. FIELD
68 Complex algebraic surfaces, A. BEAUVILLE
69 Representation theory, I.M. GELFAND *et al*
74 Symmetric designs: an algebraic approach, E.S. LANDER
76 Spectral theory of linear differential operators and comparison algebras, H.O. CORDES
77 Isolated singular points on complete intersections, E.J.N. LOOIJENGA
78 A primer on Riemann surfaces, A.F. BEARDON
79 Probability, statistics and analysis, J.F.C. KINGMAN & G.E.H. REUTER (eds)
80 Introduction to the representation theory of compact and locally compact groups, A. ROBERT
81 Skew fields, P.K. DRAXL
82 Surveys in combinatorics, E.K. LLOYD (ed)
83 Homogeneous structures on Riemannian manifolds, F. TRICERRI & L. VANHECKE
85 Solitons, P.G. DRAZIN
86 Topological topics, I.M. JAMES (ed)
87 Surveys in set theory, A.R.D. MATHIAS (ed)
88 FPF ring theory, C. FAITH & S. PAGE
89 An F-space sampler, N.J. KALTON, N.T. PECK & J.W. ROBERTS
90 Polytopes and symmetry, S.A. ROBERTSON
91 Classgroups of group rings, M.J. TAYLOR
92 Representation of rings over skew fields, A.H. SCHOFIELD
93 Aspects of topology, I.M. JAMES & E.H. KRONHEIMER (eds)
94 Representations of general linear groups, G.D. JAMES
95 Low-dimensional topology 1982, R.A. FENN (ed)
96 Diophantine equations over function fields, R.C. MASON
97 Varieties of constructive mathematics, D.S. BRIDGES & F. RICHMAN
98 Localization in Noetherian rings, A.V. JATEGAONKAR
99 Methods of differential geometry in algebraic topology, M. KAROUBI & C. LERUSTE
100 Stopping time techniques for analysts and probabilists, L. EGGHE

101 Groups and geometry, ROGER C. LYNDON
103 Surveys in combinatorics 1985, I. ANDERSON (ed)
104 Elliptic structures on 3-manifolds, C.B. THOMAS
105 A local spectral theory for closed operators, I. ERDELYI & WANG SHENGWANG
106 Syzygies, E.G. EVANS & P. GRIFFITH
107 Compactification of Siegel moduli schemes, C-L. CHAI
108 Some topics in graph theory, H.P. YAP
109 Diophantine Analysis, J. LOXTON & A. VAN DER POORTEN (eds)
110 An introduction to surreal numbers, H. GONSHOR
111 Analytical and geometric aspects of hyperbolic space, D.B.A.EPSTEIN (ed)
112 Low-dimensional topology and Kleinian groups, D.B.A. EPSTEIN (ed)
113 Lectures on the asymptotic theory of ideals, D. REES
114 Lectures on Bochner-Riesz means, K.M. DAVIS & Y-C. CHANG
115 An introduction to independence for analysts, H.G. DALES & W.H. WOODIN
116 Representations of algebras, P.J. WEBB (ed)
117 Homotopy theory, E. REES & J.D.S. JONES (eds)
118 Skew linear groups, M. SHIRVANI & B. WEHRFRITZ
119 Triangulated categories in the representation theory of finite-dimensional algebras, D. HAPPEL
121 Proceedings of *Groups - St Andrews 1985*, E. ROBERTSON & C. CAMPBELL (eds)
122 Non-classical continuum mechanics, R.J. KNOPS & A.A. LACEY (eds)
123 Surveys in combinatorics 1987, C. WHITEHEAD (ed)
124 Lie groupoids and Lie algebroids in differential geometry, K. MACKENZIE
125 Commutator theory for congruence modular varieties, R. FREESE & R. MCKENZIE
126 Van der Corput's method for exponential sums, S.W. GRAHAM & G. KOLESNIK
127 New directions in dynamical systems, T.J. BEDFORD & J.W. SWIFT (eds)
128 Descriptive set theory and the structure of sets of uniqueness, A.S. KECHRIS & A. LOUVEAU
129 The subgroup structure of the finite classical groups, P.B. KLEIDMAN & M.W.LIEBECK
130 Model theory and modules, M. PREST
131 Algebraic, extremal & metric combinatorics, M-M. DEZA, P. FRANKL & I.G. ROSENBERG (eds)
132 Whitehead groups of finite groups, ROBERT OLIVER
133 Linear algebraic monoids, MOHAN S. PUTCHA
134 Number theory and dynamical systems, M. DODSON & J. VICKERS (eds)
135 Operator algebras and applications, 1, D. EVANS & M. TAKESAKI (eds)
136 Operator algebras and applications, 2, D. EVANS & M. TAKESAKI (eds)
137 Analysis at Urbana, I, E. BERKSON, T. PECK, & J. UHL (eds)
138 Analysis at Urbana, II, E. BERKSON, T. PECK, & J. UHL (eds)
139 Advances in homotopy theory, S. SALAMON, B. STEER & W. SUTHERLAND (eds)
140 Geometric aspects of Banach spaces, E.M. PEINADOR and A. RODES (eds)
141 Surveys in combinatorics 1989, J. SIEMONS (ed)
142 The geometry of jet bundles, D.J. SAUNDERS
143 The ergodic theory of discrete groups, PETER J. NICHOLLS
144 Uniform spaces, I.M. JAMES
145 Homological questions in local algebra, JAN R. STROOKER
146 Cohen-Macaulay modules on Cohen-Macaulay rings, Y. YOSHINO
147 Continuous and discrete modules, S.H. MOHAMED & B.J. MÜLLER

London Mathematical Society Lecture Note Series. 147

Continuous and Discrete Modules

Saad H. Mohamed
Department of Mathematics, Kuwait University
Bruno J. Müller
Department of Mathematics, McMaster University

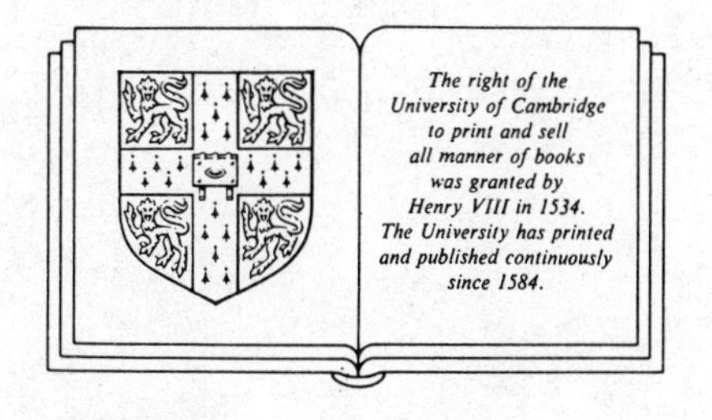

CAMBRIDGE UNIVERSITY PRESS
Cambridge
New York Port Chester Melbourne Sydney

Published by the Press Syndicate of the University of Cambridge
The Pitt Building, Trumpington Street, Cambridge CB2 1RP
40 West 20th Street, New York, NY 10011, USA
10, Stamford Road, Oakleigh, Melbourne 3166, Australia

First published 1990

Printed in Great Britain at the University Press, Cambridge

Library of Congress cataloguing in publication data available

British Library cataloguing in publication data available

ISBN 0 521 39975 0

PREFACE

The monograph addresses research mathematicians and graduate students interested in the module and representation theory of arbitary rings. It is primarily concerned with generalizations of injectivity and projectivity, and simultaneously with modules displaying good direct decomposition properties. Specifically, we study two classes of modules, named continuous and discrete. Both exhibit, in a dual sense, a generous supply of direct summands. The first class contains all injective modules, while the second one contains those projective modules which have a "good" direct sum decomposition.

Continuous, as the term is used here, is not related to continuity in the sense of topology and analysis. It is rather derived from the notion of a continuum. This usage originated with von Neumann's continuous geometries. These are analogues of projective geometries, except that they have no points, but instead a dimension function whose range is a continuum of real numbers. Just as most projective geometries can be coordinatized by simple artinian rings, most continuous geometries are coordinatized by non–noetherian continuous regular rings.

Utumi observed that continuous regular rings generalize self–injective regular rings. He extended the concept to arbitrary rings. Jeremy, Mohamed and Bouhy, and Goel and Jain generalized these ideas to modules.

The weaker notion of quasi–continuity appears now to be more fundamental. It asserts directly that the module inherits all direct sum decompositions from its injective hull (2.8). The important Theorem (2.31) ensures that uniqueness properties are inherited as well.

Another central result for quasi–continuous modules, (2.37), establishes a decomposition into a quasi–injective and a square–free part. This is a rare instance of a direct decomposition where both summands have, in different ways, better properties than the original module. It allows us to prove the exchange property for continuous modules (3.24), and the cancellation property for directly finite continuous modules (3.25).

Arbitrary families of orthogonal idempotents, of the endomorphism ring of a quasi–continuous module, lift modulo the ideal of endomorphisms with essential kernel (3.9). The endomorphism ring of a quasi–continuous module retains all the properties familiar from quasi–injective modules precisely if the module is actually continuous (3.15).

The dichotomy between projective and continuous geometries, namely that their dimension functions have discrete respectively continuum range, remains in effect for injective modules (cf. Goodearl and Boyle [76]; replace the dimension function by the finite rank function), and consequently for quasi–continuous modules. Noetherian rings are exactly the ones for which every injective or every quasi–continuous module is a direct sum of indecomposables. On the other hand, over "arbitrary" rings, the continuous structure is typical: an infinite direct sum of indecomposable (quasi–)continuous modules is (quasi–)continuous only in the presence of an ascending chain condition (2.13/3.16). A quasi–continuous module decomposes into indecomposables only in the presence of strong additional properties (2.22).

Concepts dual to those of (quasi–)continuity have been studied, under various names (notably (quasi–,semi–)perfect, (quasi–)dual continuous, stark supplementiert), by many authors. The usage of terminology is disturbingly inconsistent. We propose the new term "(quasi–)discrete", motivated by Oshiro's Theorem (4.15) that every such module is the direct sum of indecomposables.

This decomposition, which has strong uniqueness properties, reduces some proofs to counting arguments. Exchange and cancellation property, in particular, follow quite easily (4.19/20). Arbitrary families of orthogonal idempotents, of the endomorphism ring of a quasi–discrete module, lift modulo the ideal of endomorphisms with small image (5.9). Again, a quasi–discrete module is discrete precisely if the endomorphism ring exhibits all the familiar properties (5.4).

The converse question, when a direct sum of indecomposable quasi–discrete modules is quasi–discrete, has not yet received a fully satisfactory answer (cf. (4.48/49). The special cases of a finite direct sum (4.50), and a direct sum of local modules (4.53), are settled. For discrete modules over commutative noetherian rings, the complete answer is known (5.15/16), and requires elaborate arguments.

In spite of the dual nature of their definitions, and some analogies on an elementary level, (quasi–)continuous and (quasi–)discrete modules display striking dissimilarities as well: Continuity generalizes injectivity. The structure of quasi–continuous modules resembles that of their injective hulls. All injective/quasi–continuous modules are direct sums of indecomposables if and only if the ring is noetherian. A direct sum of quasi–continuous modules with full relative injectivity is quasi–continuous. On the other hand: Quasi–discrete modules are always direct sums of indecomposables. Discreteness generalizes projectivity if and only if the ring is perfect. A direct sum of quasi–discrete modules with full relative projectivity need not be quasi–discrete.

These dissimiliarities can be traced to the fact that every module is complemented, and hence possesses an injective hull, while most modules are not supplemented (4.41), and have no projective cover. Modules which are just supplemented, have interesting properties of their own, and were studied by many people (cf. Appendix, Section 1 and 2). Quasi–discrete modules are supplemented (4.8), and thus constitute a much more restrictive class than quasi–continuous ones.

Chapter 1 is of preliminary nature. It summarizes facts about relative injectivity, and proves the exchange and cancellation properties for injective modules. It also develops a general technique for constructing direct sum decompositions, and derives the decomposition of an injective module into a directly finite and a purely infinite part. Analogous results on relative projectivity are collected at the beginning of Section 4 of Chapter 4. More details concerning the arrangement of the material may be obtained from the table of content.

We have attempted to provide a complete and up to date account of the subject. The exposition is self contained, except that a few well known and highly technical results which are readily accessible in the literature, are quoted without proof. All undefined concepts can be found in Anderson and Fuller [73]. In the comments at the end of each chapter, we try to trace the origin of some of the main ideas. Section 6 of the Appendix lists a number of open questions.

M M ☺ H H

The main portion of this work was done while the first author was on sabbatical leave from Kuwait University, at the second author's institution, McMaster University. The authors acknowledge partial financial support by Kuwait University and by the National Science and Engineering Research Council of Canada.

Saad H. Mohamed

Bruno J. Müller

TABLE OF CONTENTS

CHAPTER 1

INJECTIVITY AND RELATED CONCEPTS

In this chapter we discuss injectivity, quasi–injectivity and relative injectivity, with emphasis on those properties which are used later on in the book. We start by listing some of the well known fundamental properties of injective modules which can be found in Anderson and Fuller [73] or Sharpe and Vamos [72].

A module E is *injective* if it satisfies any of the equivalent conditions:

(1) For every module A and any submodule X of A every homomorphism X ⟶> E can be extended to a homomorphism A ⟶> E;

(2) (Baer's Criterion) Every homomorphism of a right ideal I of R to E can be extended to a homomorphism of R to E;

(3) For any module M every monomorphism E >⟶> M splits;

(4) E has no proper essential extensions.

Every module M has a minimal injective extension, which is at the same time a maximal essential extension of M; such an extension is unique up to isomorphism and is called the *injective hull* of M. The injective hull of M will be denoted by E(M).

1. A–INJECTIVE MODULES

__Definition 1.1__. Let A be an R–module. A module N is said to be A–*injective* if for every submodule X of A, any homomorphism φ: X ⟶> N can be extended to a homomorphism ψ : A ⟶> N.

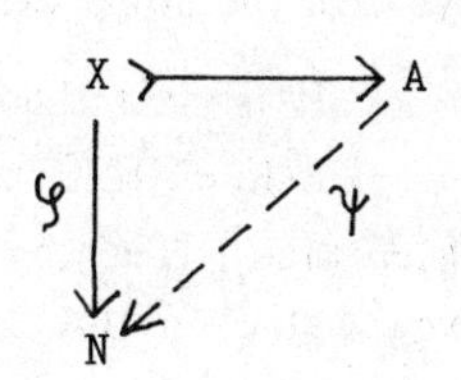

The following is an immediate consequence.

__Lemma 1.2__. *If* N *is* A*–injective, then any monomorphism* N $>\xrightarrow{f}>$ A *splits. If, in addition,* A *is indecomposable, then* f *is an isomorphism.* □

__Proposition 1.3__. *Let* N *be an* A*–injective module. If* B ≤ A, *then* N *is* B*–injective and* A/B*–injective.*

PROOF. It is obvious that N is B–injective. Let X/B be a submodule of A/B, and $\varphi : X/B \longrightarrow N$ be a homomorphism. Let π denote the natural homomorphism of A onto A/B and $\pi' = \pi|_X$. Since N is A–injective, there exists a homomorphism $\theta : A \longrightarrow N$ that extends $\varphi\pi'$. Now

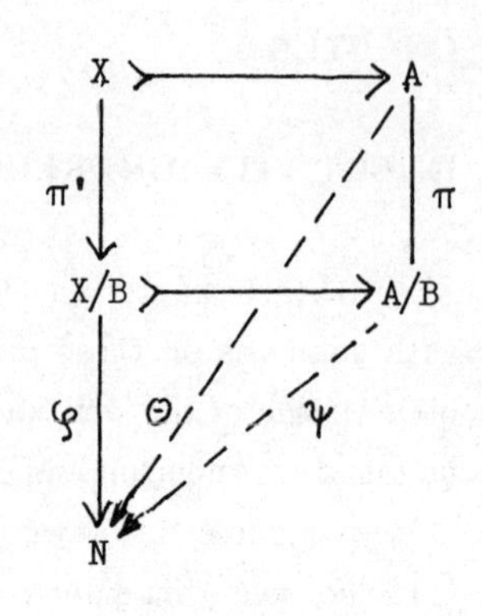

$$\theta B = \varphi\pi' B = \varphi(0) = 0.$$

Hence Ker $\pi \leq$ Ker θ, and consequently there exists $\psi : A/B \longrightarrow N$ such that $\psi\pi = \theta$. For every $x \in X$

$$\psi(x + B) = \psi\pi(x) = \theta(x) = \varphi\pi'(x) = \varphi(x + B).$$

Thus ψ extends φ, and therefore N is A/B injective. □

The following proposition may be viewed as a generalization of Baer's Criterion.

Proposition 1.4. *A module* N *is A–injective if and only if* N *is* aR*–injective for every* $a \in A$.

PROOF. The "only if" part follows by the preceding proposition.

Conversely, assume that N is aR–injective for every $a \in A$. Let $X \leq A$ and $\varphi : X \longrightarrow N$ be a homomorphism. By Zorn's Lemma, we can find a pair (B, ψ) maximal with the properties $X \leq B \leq A$ and $\psi : B \longrightarrow N$ is a homomorphism which extends φ. It is clear that $B \leq^e A$. Suppose that $B \neq A$ and consider an element $a \in A - B$. Let $K = \{r \in R : ar \in B\}$; then it is clear that $aK \neq 0$. Define $\mu : aK \longrightarrow N$ by $\mu(ak) = \psi(ak)$. Then by assumption μ can be extended to $\nu : aR \longrightarrow N$.

Now define $\chi : B + aR \longrightarrow N$ by $\chi(b + ar) = \psi(b) + \nu(ar)$. Then χ is well defined, since if $b + ar = 0$, then $r \in K$ and so

$$\psi(b) + \nu(ar) = \psi(b) + \mu(ar) = \psi(b) + \psi(ar) = \psi(b + ar) = 0.$$

But then the pair (B + aR, χ) contradicts the maximality of (B, ψ). Hence B = A, and $\psi : A \longrightarrow N$ extends φ. □

Proposition 1.5. *A module* N *is* $(\bigoplus_{i \in I} A_i)$*–injective if and only if* N *is* A_i*–injective for every* $i \in I$.

PROOF. Assume that N is A_i–injective for all $i \in I$. Let $A = \bigoplus_{i \in I} A_i$, $X \leq A$ and consider a homomorphism $\varphi : X \longrightarrow N$. We may assume, by Zorn's Lemma, that φ cannot be extended to a homomorphism $X' \longrightarrow N$ for any submodule X′ of A which contains X properly. Then $X \leq^e A$. We claim that X = A. Suppose not. Then there exist $j\in I$ and $a \in A_j$ such that $a \notin X$. Since N is A_j–injective, N is aR–injective by Proposition 1.3. By an argument similar to that given in Proposition 1.4, we can extended φ to a homomorphism ψ : $X + aR \longrightarrow N$, which contradicts the maximality of φ. This proves our claim, and hence N is A–injective.

The converse follows by Proposition 1.3. □

The same proof as for injective modules yields the following

<u>**Proposition 1.6**</u>. $\prod_{\alpha\in\Lambda} M_\alpha$ *is* A–*injective if and only if* M_α *is* A–*injective for every* $\alpha\in\Lambda$. □

Next we investigate the A–injectivity of direct sums.

<u>**Theorem 1.7**</u>. *The following are equivalent for a family of modules* $\{M_\alpha : \alpha\in\Lambda\}$:

(1) $\bigoplus_{\alpha\in\Lambda} M_\alpha$ *is* A–*injective*;

(2) $\bigoplus_{i\in I} M_i$ *is* A–*injective for every countable subset* $I \subseteq \Lambda$;

(3) M_α *is* A–*injective for every* $\alpha\in\Lambda$, *and for every choice of* $m_i\in M_{\alpha_i}$ $(i\in\mathbb{N})$ *for distinct* $\alpha_i\in\Lambda$ *such that* $\bigcap_{i=1}^{\infty} m_i^o \geq a^o$ *for some* $a\in A$, *the ascending sequence* $\bigcap_{i\geq n} m_i^o$ $(n\in\mathbb{N})$ *becomes stationary.*

PROOF. (1) ⇒ (2) follows by Proposition 1.6.

(2) ⇒ (3): Proposition 1.6 implies that M_α is a A–injective for every $\alpha\in A$. Consider the element $x = (m_i) \in \prod_{i=1}^{\infty} M_{\alpha_i}$. The mapping φ: $ar \to xr$ is a well defined homomorphism from aR to $\prod_{i=1}^{\infty} M_{\alpha_i}$. Let $I = \bigcup_{n=1}^{\infty} (\bigcap_{i\geq n} m_i^o)$, and let $\overline{\varphi}$ denote the restriction of φ to aI. Then $\overline{\varphi}$ is a homomorphism of aI into $\bigoplus_{i=1}^{\infty} M_{\alpha_i}$. Since $\bigoplus_{i=1}^{\infty} M_{\alpha_i}$ is A–injective and hence aR–injective, $\overline{\varphi}$ extends to ψ: $aR \longrightarrow \bigoplus_{i=1}^{\infty} M_{\alpha_i}$. Then

$$xI = \overline{\varphi}(aI) = \psi(aI) \leq \psi(aR) = \psi(a)R \leq \bigoplus_{i\in F} M_{\alpha_i},$$

where F is a finite subset of $\mathbb{N}$. Let $F = \{1,2,...,k-1\}$. Then $m_iI = 0$ for $i \geq k$ and hence $I = \bigcap_{i\geq k} m_i^o$. Therefore the sequence $\bigcap_{i\geq n} m_i^o$ $(n\in\mathbb{N})$ becomes stationary.

(3) ⇒ (1): By way of contradiction, assume that $\bigoplus_{\alpha\in\Lambda} M_\alpha$ is not A–injective. Then by

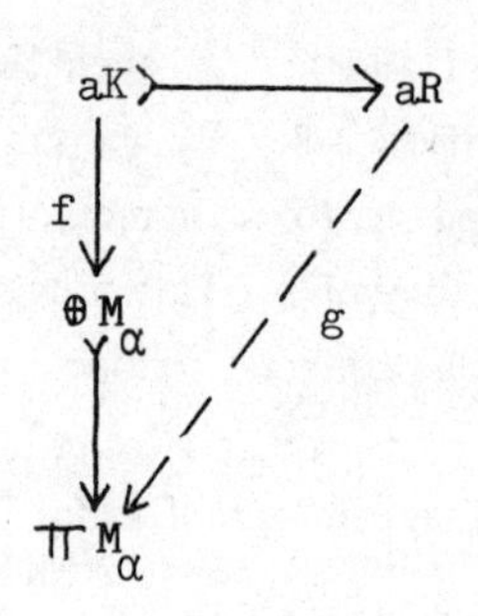

Proposition 1.4, $\bigoplus_{\alpha\in\Lambda} M_\alpha$ is not aR–injective for some a∈A. Hence there exists a right ideal K of R and a homomorphism f: aK $\longrightarrow \bigoplus_{\alpha\in\Lambda} M_\alpha$ such that f cannot be extended to aR. Since $\bigoplus_{\alpha\in F} M_\alpha$ is A–injective for all finite subsets $F \subseteq \Lambda$ by Proposition 1.6, $f(aK) \nsubseteq \bigoplus_{\alpha\in F} M_\alpha$ for any finite subset $F \subseteq \Lambda$. However f can be extended to $g : aR \longrightarrow \prod_{\alpha\in\Lambda} M_\alpha$ since $\prod_{\alpha\in\Lambda} M_\alpha$ is A–injective. Let $m = g(a)$. Then it is clear that $a^o \leq m^o = \bigcap_{\alpha\in\Lambda} m_\alpha^o$ where m_α is the α–component of $m\in \prod_{\alpha\in\Lambda} M_\alpha$. Then Let $S_k = \{\alpha\in\Lambda : m_\alpha k \neq 0\}$, $k\in K$. Then S_k is a finite subset of Λ for every $k\in K$. However $I = \bigcup_{k\in K} S_k$ is not finite since $mK = f(aK) \nsubseteq \bigoplus_{\alpha\in F} M_\alpha$ for any finite subset $F \subseteq \Lambda$. By induction we select elements $k_i\in K$ ($i\in\mathbb{N}$) and indices $\alpha_j\in\Lambda$ such that $\alpha_j\in S_{k_j}$ and $\alpha_j \notin \bigcup_{i=1}^{j-1} S_{k_i}$. Let m_i denote the α_i–component of m. Then $a^o \leq \bigcap_{i=1}^{\infty} m_i^o$ and the sequence $\bigcap_{i\geq n} m_i^o$ ($n\in\mathbb{N}$) is strictly increasing, which is a contraction to our assumption. Therefore $\bigoplus_{\alpha\in\Lambda} M_\alpha$ is A–injective. □

<u>Corollary 1.8</u>. $\bigoplus_{i=1}^{\infty} M_i$ *is* A*–injective if and only if* M_i *is* A*–injective for every* $i\in\mathbb{N}$, *and for every choice* $m_i\in M_i$ *such that* $\bigcap_{i=1}^{\infty} m_i^o \geq a^o$ *for some* a∈A, *the ascending sequence* $\bigcap_{i\geq n} m_i^o$ ($n\in\mathbb{N}$) *becomes stationary.* □

Motivated by these results and later applications, we introduce the following three chain conditions on a ring R relative to a given family of R–modules $\{M_\alpha : \alpha\in\Lambda\}$:

(A_1) For every choice of distinct $\alpha_i\in\Lambda$ ($i\in\mathbb{N}$) and $m_i\in M_{\alpha_i}$ the ascending sequence $\bigcap_{i\geq n} m_i^o$ ($n\in\mathbb{N}$) becomes stationary;

(A_2) Fore every choice of $x\in M_\alpha$ ($\alpha\in\Lambda$) and $m_i\in M_{\alpha_i}$ for distinct $\alpha_i\in\Lambda$ ($i\in\mathbb{N}$) such that $m_i^o \geq x^o$, the ascending sequence $\bigcap_{i\geq n} m_i^o$ ($n\in\mathbb{N}$) becomes stationary;

(A_3) For every choice of distinct $\alpha_i \in \Lambda$ $(i \in \mathbb{N})$ and $m_i \in M_{\alpha_i}$, if the sequence m_i^o is ascending, then it becomes stationary.

It is clear that (A_1) implies (A_2) and (A_2) implies (A_3). No other implication holds as we shall see by the end of this section.

The following is a consequence of Proposition 1.6 and Theorem 1.7.

<u>Proposition 1.9</u>. *Let* $M = \bigoplus_{\alpha \in \Lambda} M_\alpha$. *Then* $M(\Lambda - \alpha)$ *is* M_α*–injective for every* $\alpha \in \Lambda$ *if and only if* M_α *is* M_β*–injective for all* $\alpha \neq \beta \in \Lambda$ *and* (A_2) *holds.*

□

By Proposition 1.6, a direct product of injective modules is injective, and hence a finite direct sum of injective modules is injective. The following proposition, which deals with injectivity of arbitrary direct sums, is an immediate consequence of Theorem 1.7.

<u>Proposition 1.10</u>. $\bigoplus_{\alpha \in \Lambda} M_\alpha$ *is injective if and only if each* M_α *is injective and* (A_1) *holds.*

□

<u>Theorem 1.11</u>. *The direct sum of any family of* A*–injective modules is* A*–injective if and only if every cyclic (or finitely generated) submodule of* A *is noetherian. In particular, the direct sum of every family of injective* R*–modules is injective if and only if* R *is right noetherian.*

PROOF. Assume that aR is noetherian for every $a \in A$, and consider a direct sum $M = \bigoplus_{\alpha \in \Lambda} M_\alpha$ of A–injective modules M_α. Let $B \leq aR$ and $\varphi : B \longrightarrow M$ be a homomorphism. Since B is finitely generated, $\varphi(B) \leq \bigoplus_{\alpha \in F} M_\alpha$ for a finite subset $F \subseteq \Lambda$. Then φ can be extended to $\psi : aR \longrightarrow \bigoplus_{\alpha \in F} M_\alpha$, since $\bigoplus_{\alpha \in F} M_\alpha$ is A–injective by Proposition 1.6. Hence the A–injectivity of M follows by Proposition 1.4.

Conversely, assume that the direct sum of any family of A–injective modules is A–injective. Let a be an arbitrary element of A. We prove that aR is right noetherian by showing that any ascending sequence

$$a^o = B_o \leq B_1 \leq B_2 \leq \ldots$$

of right ideals of R is ultimately stationary. Let $M_i = E(R/B_i)$, $i \in \mathbb{N}$. Since each M_i is trivially A–injective, $\bigoplus_{i=1}^{\infty} M_i$ is A–injective by assumption. Consider the set of elements $\{m_i = 1 + B_i : i \in \mathbb{N}\}$. The A–injectivity of $\bigoplus_{i=1}^{\infty} M_i$ implies, by Corollary 1.8, that the ascending sequence $\bigcap_{i \geq n} m_i^o$ $(n \in \mathbb{N})$ becomes stationary. As $m_i^o = B_i$ for every $i \in \mathbb{N}$,

$$B_n = m_n^o = \bigcap_{i \geq n} m_i^o.$$

Hence the sequence $B_1 \leq B_2 \leq \ldots$ becomes stationary, and consequently aR is noetherian.

The last statement is obvious. □

We conclude this section by listing examples which seperate the ascending chain conditions (A_1), (A_2) and (A_3). Each of these examples is of the type $\bigoplus_{i \in \mathbb{N}} M_i$ with indecomposable injective M_i.

<u>Examples 1.12</u>. (1) Let R be any commutative domain, and let K be its quotient field. If we take $M_i = K$ $(i \in \mathbb{N})$, then $\bigoplus_{i \in \mathbb{N}} M_i$ is injective, hence (A_1) holds. However R is not necessarily noetherian.

(2) Let $R = \prod_{i \in \mathbb{N}} K_i$, a product of fields; and $M_i = K_i$. Here $\bigoplus_{i \in \mathbb{N}} M_i$ is semisimple, hence it is obvious that (A_2) holds. Since $E(\bigoplus_{i \in \mathbb{N}} M_i) = \prod_{i \in \mathbb{N}} K_i$, $\bigoplus_{i \in \mathbb{N}} M_i$ is not injective, hence (A_1) does not hold.

(3) Let R be any (left and right) perfect ring such that $E(R_R)$ is projective but $E({}_RR)$ is not (for the existence of such a ring, see Müller [68]). Let M be a direct sum of countably many copies of $E(R_R)$. Then M is not quasi–injective by (Yamagata [74], Lemma 3.1). Since $E(R_R)$ is projective, it is a finite direct sum of indecomposables; so $M = \bigoplus_{i \in \mathbb{N}} M_i$, with each M_i indecomposable injective. That $\bigoplus_{i \in \mathbb{N}} M_i$ does not have (A_2) follows by Proposition 1.9 (see also Proposition 1.18). - Proposition 2.24 (see Definition 2.23).

(4) For an incidence where even (A_3) fails, consider any local generalized quasi–Frobenius ring, and let $M_i = R (i \in \mathbb{N})$. Then $\bigoplus_{i \in \mathbb{N}} M_i$ has (A_3) if and only if it is locally–semi–T–nilpotent (Proposition 2.24), consequently R is perfect and hence quasi–Frobenius. An explicit example of a local generalized quasi–Frobenius ring which is not quasi–Frobenius is $R = \hat{\mathbb{Z}}_p \ltimes C_p^{\infty}$, the split extension of the ring $\hat{\mathbb{Z}}_p$ of p–adic integers by the Prüfer group C_p^{∞}.

2. QUASI–INJECTIVE MODULES

A module Q is called *quasi–injective* if it is Q–injective. Quasi–injective modules are closely related to their injective hulls. We investigate this relationship in a more general setting.

Lemma 1.13. *A module* N *is* A*-injective if and only if* $\psi A \leq N$ *for every* $\psi \in \mathrm{Hom}\,(E(A), E(N))$.

PROOF. Since E(N) is injective, it is enough to consider $\psi \in \mathrm{Hom}\,(A, E(N))$.

"If": Let $X \leq A$ and $\varphi : X \longrightarrow N$ be a homomorphism. Since E(N) is injective, φ can be extended to $\psi : A \longrightarrow E(N)$. By assumption $\psi A \leq N$, and hence $\psi : A \longrightarrow N$ extends φ. Therefore N is A–injective.

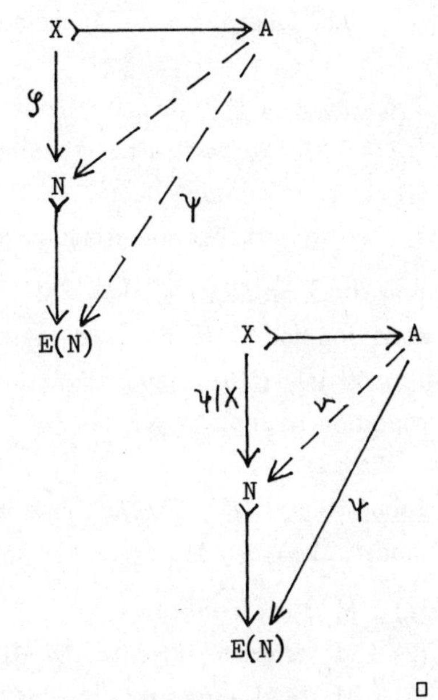

"Only if" : Let $X = \{a \in A : \psi(a) \in N\}$. Since N is A–injective, $\psi|_X$ can be extended to $\nu : A \longrightarrow N$. We claim that $N \cap (\nu - \psi)A = 0$. Indeed, let $n \in N$ and $a \in A$ be such that $n = (\nu - \psi)(a)$. Then $\psi(a) = \nu(a) - n \in N$, and consequently $a \in X$. Then $n = \nu(a) - \psi(a) = \psi(a) - \psi(a) = 0$.

Therefore $N \cap (\nu - \psi)A = 0$, and hence $(\nu - \psi)A = 0$ as $N \leq^e E(N)$. Hence $\psi A = \nu A \leq N$. □

Corollary 1.14. *A module* Q *is quasi–injective if and only if* $fQ \leq Q$ *for every* $f \in \mathrm{End}\,E(Q)$. □

Corollary 1.15. *Every module* M *has a minimal quasi–injective extension, which is unique up to isomorphism.*

PROOF. Let $Q(M) = (\mathrm{End}\,E(M))(M)$. Then it is obvious that Q(M) satisfies the required conditions. □

Lemma 1.13 has also the following

Corollary 1.16. *Let* A *and* B *be relatively injective* (i.e. A is B–injective and B is A–injective). *If* $E(A) \cong E(B)$, *then* $A \cong B$; *in fact any isomorphism* $E(A) \longrightarrow E(B)$ *restricts to an isomorphism* $A \longrightarrow B$; *in addition* A *and* B *are quasi–injective.*

PROOF. Let $g : E(A) \longrightarrow E(B)$ be an isomorphism. Since B is A–injective, $gA \leq B$ by Lemma 1.13. Similarly $g^{-1}B \leq A$. Hence

$$B = (gg^{-1})B = g(g^{-1}B) \leq gA \leq B.$$

Consequently $gA = B$, and therefore $g|_A : A \longrightarrow B$ is an isomorphism.

Since A is B–injective and $B \cong A$, A is A–injective, that is A is quasi–injective.

□

The following is an immediate consequence of Propositions 1.5 and 1.6.

<u>Proposition 1.17</u>. $M_1 \oplus M_2$ *is quasi–injective if and only if* M_i *is* M_j*–injective* $(i,j = 1,2)$. *In particular, a summand of a quasi–injective module is quasi–injective.*

□

Now consider an arbitrary direct sum $M = \bigoplus_{\alpha\in\Lambda} M_\alpha$. It is clear from the preceding proposition that "M_α is M_β–injective for all $\alpha, \beta\in\Lambda$" is a necessary condition for M to be quasi–injective. The following result (which is analogous to Proposition 1.10) shows that this condition is also sufficient in the presence of condition (A_2).

<u>Proposition 1.18</u>. *The following are equivalent for a direct sum decomposition of a module* $M = \bigoplus_{\alpha\in\Lambda} M_\alpha$:

(1) M *is quasi–injective;*

(2) M_α *is quasi–injective and* $M(\Lambda - \alpha)$ *is* M_α*–injective for every* $\alpha\in\Lambda$;

(3) M_α *is* M_β*–injective for all* $\alpha, \beta\in\Lambda$ *and* (A_2) *holds.*

PROOF. Using Propositions 1.9 and 1.17, the proof is straightforward. □

<u>Corollary 1.19</u>. $\bigoplus_{i=1}^{n} M_i$ *is quasi–injective if and only if* M_i *is* M_j*–injective* $(i, j = 1,2,\ldots,n)$. M^n *is quasi–injective if and only if* M *is quasi–injective.*

□

3. EXCHANGE AND CANCELLATION PROPERTIES

In this section, we prove that every (quasi–)injective module has the exchange property, and an injective module has the cancellation property if and only if it is directly finite. (The terms are defined below).

Definition 1.20. A module M is said to have the *(finite) exchange property* if for any (finite) index set I, whenever $M \oplus N = \bigoplus_{i \in I} A_i$ for modules N and A_i, then $M \oplus N = M \oplus (\bigoplus_{i \in I} B_i)$ for submodules $B_i \leq A_i$.

It is fairly easy to show that the (finite) exchange property is inherited by summands and finite direct sums (see Lemma 3.20).

Theorem 1.21. *Every quasi-injective module* M *has the exchange property.*

PROOF. Let $A = M \oplus N = \bigoplus_{i \in I} A_i$. Let $X_i = A_i \cap N$ and $X = \bigoplus_{i \in I} X_i$. By Zorn's Lemma, we can find $B \leq A$ maximal with respect to the following properties:

(i) $B = \bigoplus_{i \in I} B_i$ with $X_i \leq B_i \leq A_i$,

(ii) $M \cap B = 0$.

We claim that $A = M \oplus B$. Our claim will hold if we show that $\overline{M} \leq^e \overline{A}$ and $\overline{M} \subset^{\oplus} \overline{A}$ (where $\overline{Y}$ denotes the image of $Y \leq A$ under the natural homomorphism $A \longrightarrow A/B$). We start by showing $\overline{M} \cap \overline{A}_j \leq^e \overline{A}_j$ for every $j \in I$. Let D be an arbitrary submodule of A_j such that $B_j < D$. Then

$$B < D + B = D \oplus (\bigoplus_{i \neq j} B_i).$$

Maximality of B then implies $M \cap (D + B) \neq 0$. Since $M \cap B = 0$, $M \cap (D + B) \nleq B$. Hence

$$(\overline{M} \cap \overline{A}_j) \cap \overline{D} = \overline{M} \cap \overline{D} \neq 0.$$

Hence $\overline{M} \cap \overline{A}_j \leq^e \overline{A}_j$ for all $j \in I$. Consequently

$$\bigoplus_{j \in I} (\overline{M} \cap \overline{A}_j) \leq^e \bigoplus_{j \in I} \overline{A}_j = \overline{A}.$$

Therefore $\overline{M} \leq^e \overline{A}$.

If M is injective, then $\overline{M}$ is also injective (since $\overline{M} = (M \oplus B)/B \cong M$) and $\overline{M} \subset^{\oplus} \overline{A}$ follows trivially. For a quasi- injective module we have the following additional argument: Let π be the projection $M \oplus N \longrightarrow\!\!\!> M$. The restriction of π to A_i has kernel X_i, and hence A_i/X_i is isomorphic to a submodule of M. Since M is M–injective, M is A_i/X_i–injective by Proposition 1.3. As $A/X \cong \oplus A_i/X_i$, we get by Proposition 1.5 that M is A/X–injective; hence M is A/B–injective by Proposition 1.3. Since $\overline{M} \cong M$, $\overline{M}$ is $\overline{A}$–injective and therefore $\overline{M} \subset^{\oplus} \overline{A}$ by Lemma 1.2. □

Definition 1.22. A module M is said to have the *cancellation property* if whenever $M \oplus X \cong M \oplus Y$, then $X \cong Y$. M is said to have the *internal cancellation property* if whenever $M = A_1 \oplus B_1 = A_2 \oplus B_2$ with $A_1 \cong A_2$, then $B_1 \cong B_2$.

<u>Proposition 1.23</u>. *Let* M *be a module with the finite exchange property. Then* M *has the cancellation property if and only if* M *has the internal cancellation property.*

PROOF. "Only if" : Let $M = A_1 \oplus B_1 = A_2 \oplus B_2$ with $A_1 \cong A_2$. Then

$$M \oplus B_1 = A_2 \oplus B_2 \oplus B_1 \cong A_1 \oplus B_2 \oplus B_1 = M \oplus B_2.$$

Hence $B_1 \cong B_2$. (This direction of the proof does not need the finite exchange property on M).

"If" : Let $M \oplus X = N \oplus Y$ with $M \cong N$. By the finite exchange property we get $M \oplus X = M \oplus N' \oplus Y'$ such that $N' \leq N$ and $Y' \leq Y$. Then $X \cong N' \oplus Y'$. It is also clear that $N' \subset^{\oplus} N$ and $Y' \subset^{\oplus} Y$; write $N = N' \oplus N''$ and $Y = Y' \oplus Y''$. Then

$$M \oplus N' \oplus Y' = M \oplus X = N \oplus Y = N' \oplus N'' \oplus Y' \oplus Y''.$$

Hence $M \cong N'' \oplus Y''$, and therefore

$$N'' \oplus Y'' \cong M \cong N = N'' \oplus N'.$$

Since M has the internal cancellation property, $N' \cong Y''$; hence

$$X \cong N' \oplus Y' \cong Y'' \oplus Y' = Y. \qquad \square$$

<u>Definition 1.24</u>. A module D is called *directly finite* if D is not isomorphic to a proper summand of itself.

It is clear that a summand of a directly finite module is again directly finite.

The following is a characterization of directly finite modules via their endomorphism rings.

<u>Proposition 1.25</u>. *A module* D *is directly finite if and only if* $fg = 1$ *implies that* $gf = 1$ *for all* $f, g \in \text{End } M$.

PROOF. Assume that $fg = 1$ for some $f, g \in \text{End } M$. Then $D = gD \oplus \text{Ker} f$. Since $gD \cong D$ and D is directly finite, $\text{Ker} f = 0$. However f is onto, and therefore f is an automorphism of D. Hence $gf = 1$.

Conversely, assume the condition and let $D = B \oplus C$ with $B \cong D$. Let $\varphi : D \longrightarrow B$ be an isomorphism. Define φ^* as φ^{-1} on B and 0 on C. Then $\varphi^* \varphi = 1$ and hence $\varphi \varphi^* = 1$. It then follows that φ^* is a monomorphism and hence $C = 0$.

$\square$

<u>Lemma 1.26</u>. *If* M *is not directly finite, then* $X^{(\mathbb{N})}$ *embeds in* M *for some non-zero module* X.

PROOF. Since M is not directly finite, $M = A \oplus X$ with $A \cong M$ and $X \neq 0$. Then $A = A_1 \oplus X_1$ with $A_1 \cong A$ and $X_1 \cong X$. Iterating this process we get

$$A = A_n \oplus X_n \oplus \dots \oplus X_2 \oplus X_1$$

with $X_i \cong X$ ($n \in \mathbb{N}$). Hence M contains the infinite direct sum $\bigoplus_{i=1}^{\infty} X_i$ with $X_i \cong X$.

□

For injective modules, we show that this condition is also sufficient.

<u>Proposition 1.27</u>. *An injective module* M *is not directly finite if and only if* $X^{(\mathbb{N})}$ *embeds in* M *for some non-zero module* X.

PROOF. In view of Lemma 1.26, we need only prove the "if" part. Assume that $M \geq K = \bigoplus_{i=1}^{\infty} X_i$ with $X_i \cong X$ and $X \neq 0$. Let $K_1 = X_1$ and $K = \bigoplus_{i=2}^{\infty} X_i$. Then $K = K_1 \oplus K_2$ with $K_2 \cong K$. Consequently

Hence $E(K)$ is not directly finite. Since $E(K) \subset^{\oplus} M$, M is not directly finite.

□

<u>Proposition 1.28</u>. *A directly finite injective module* M *has the internal cancellation property.*

PROOF. Let $M = A \oplus C = B \oplus D$ with $A \cong B$. Using Zorn's Lemma we can find a monomorphism $C \geq C' \xrightarrow{f} D$ which cannot be further extended. Injectivity of C implies that $C' \subset^{\oplus} C$ and $D' = fC' \subset^{\oplus} D$. Write $C = C' \oplus C_o$, $D = D' \oplus D_o$, $A_o = A \oplus C'$ and $B_o = B \oplus D'$. Then

$$M = A_o \oplus C_o = B_o \oplus D_o \text{ with } A_o \cong B_o.$$

If $C_o = 0$, then

$$M = B_o \oplus D_o \cong A_o \oplus D_o \cong M \oplus D_o.$$

Since M is directly finite, $D_o = 0$ and hence $C \cong D$. Thus our proof will be complete if we show $C_o = 0$.

Assume that $C_o \neq 0$. As f cannot be extended, C_o and D_o have no non–zero isomorphic submodules; hence $C_o \cap D_o = 0$ and therefore $C_o \oplus D_o \subset^{\oplus} M$. Write $M = K_o \oplus C_o \oplus D_o$. Then

$$K_o \oplus D_o \cong A_o \cong B_o \cong K_o \oplus C_o.$$

Consequently

$$A_o = A_1 \oplus C_1 = B_1 \oplus D_1$$

with $A_1 \cong K_o \cong B_1$, $C_1 \cong C_o$ and $D_1 \cong D_o$. Thus $C_1 \cap D_1 = 0$ as C_o and D_o have no non–zero isomorphic submodules. The same argument applied to A_o yields

$$A_1 = A_2 \oplus C_2 = B_2 \oplus D_2$$

with $A_2 \cong B_2$, $C_2 \cong C_1$, $D_2 \cong D_1$ and $C_2 \cap D_2 = 0$. Iterating this process we get for every $n \in \mathbb{N}$

$$A_{n-1} = A_n \oplus C_n = B_n \oplus D_n$$

with $A_n \cong B_n$, $C_n \cong C$, $D_n \cong D$ and $C_n \cap D_n = 0$. It is clear that $A_n \neq 0$; otherwise $B_n = 0$ and $C_n = D_n$ in contradiction to $C_n \cap D_n = 0$. Now

$M = A_o \oplus C_o = A_1 \oplus C_1 \oplus C_o = \ldots = A_n \oplus (C_n \oplus \ldots \oplus C_1 \oplus C_o)$. This proves that M contains the direct sum $\bigoplus_{i=1}^{\infty} C_i$ with $C_i \cong C_o$, which is a contradiction to Proposition 1.27. Hence $C_o = 0$. □

__Theorem 1.29__. *An injective module* M *has the cancellation property if and only if* M *is directly finite.*

PROOF. We first note that any module M with the internal cancellation property is directly finite. Indeed, if $M = A \oplus B$ with $A \cong M$, then $M \oplus B \cong A \oplus B = M = M \oplus 0$, and hence $B = 0$.

The converse direction follows by Theorem 1.21 and Propositions 1.23 and 1.28.

□

__Remark__. The conclusion of Theorem 1.29 holds for quasi-injective modules and even for continuous modules (see Corollary 3.25).

4. DECOMPOSITION THEOREMS

In this section we shall obtain a decomposition of injective modules into directly finite and purely infinite submodules with strong uniqueness properties.

__Definitions 1.30__. Let A and B be summands of a modules M. A is said to be *perspective* to B if there exists $X \leq M$ such that $M = A \oplus X = B \oplus X$. We say that A is *superspective* to B if for any submodule $X \leq M$, $M = A \oplus X$ if and only if $M = B \oplus X$.

Perspectivity is reflexive and symmetric but not, in general, transitive; while superspectivity is an equivalence relation.

A decomposition $M = M_1 \oplus M_2$ with certain properties is said to be unique (unique up to superspectivity , unique up to isomorphism), if for any other decomposition $M = N_1 \oplus N_2$ with the same properties, $M_i = N_i$ (M_i is superspective to N_i , $M_i \cong N_i$), ($i = 1,2$).

Obviously one has the implications

unique ⇒ unique up to superspectivity ⇒ unique up to isomorphism.

<u>Lemma 1.31</u>. *Let* N *and* $\bigoplus_{i\in I} X_i$ *be submodules of a module* M. *If* $N \cap (\bigoplus_{i\in I} X_i) \neq 0$, *then there exists* $j\in I$ *such that* X_j *and* N *have non–zero isomorphic submodules.*

PROOF. $N \cap (\bigoplus_{i\in I} X_i) \neq 0$, implies $N \cap (\bigoplus_{i\in F} X_i) \neq 0$ for a finite subset $F \subseteq I$. Let K be a maximal subset of F such that $N \cap (\bigoplus_{i\in K} X_i) = 0$. Consider $j\in F-K$ and let π be the projection $X_j \oplus (\bigoplus_{i\in K} X_i) \longrightarrow\!\!\!\!\!\!\!\rightarrow X_j$. Then $N' = \cap (X_j \oplus (\bigoplus_{i\in K} X_i)) \neq 0$ and $N \geq N' \cong \pi N' \leq X_j$. □

As a tool for the following proofs, and for later applications, we describe a general method for constructing direct decompositions. The easy verifications are left to the reader.

Two modules will be called *orthogonal* if they have no non–zero isomorphic submodules. For any class $\mathscr{X}$ of modules, $\mathscr{X}^{\perp}$ denotes the class of modules orthogonal to all members of $\mathscr{X}$. It is clear that $\mathscr{X} \subseteq \mathscr{X}^{\perp\perp}$ and $\mathscr{X}^{\perp} = \mathscr{X}^{\perp\perp\perp}$.

$\mathscr{X} = \mathscr{X}^{\perp\perp}$ holds if and only if $\mathscr{X}$ is closed under isomorphisms, submodules, essential extensions and direct sums (use Lemma 1.31). Such a class $\mathscr{X}$ is also closed under extensions, and factors modulo closed submodules; but need not be closed under arbitrary factors nor products (cf. the class of torison–free and torsion abelian groups, respectively).

$\mathscr{X}^{\perp}$ and $\mathscr{X}^{\perp\perp}$ form what we call an *orthogonal pair*, i.e. a pair of classes $\mathscr{A}$ and $\mathscr{B}$ such that $\mathscr{A}^{\perp} = \mathscr{B}$ and $\mathscr{B}^{\perp} = \mathscr{A}$. Given such a pair, and an arbitrary module M, then there are submdoules A and B of M maximal with respect to $A \in \mathscr{A}$ and $B \in \mathscr{B}$. The sum A + B is direct and essential in M. A and B are not necessarily unique, not even up to isomorphism; in fact the possible B's are precisely the complements in M of any A and vice versa.

If M is injective (or more generally quasi–continuous; cf. Lemma 2.36), we obtain a direct decomposition $M = A \oplus B$, which is unique up to superspectivity.

If ones starts (as we will in the applications) with a *hereditary* class, i.e. a class $\mathscr{X}$ closed under isomorphisms and submodules, then

$\mathscr{X}^{\perp} = \{V : V \text{ has no non–zero submodule in } \mathscr{X}\}$

$\mathscr{X}^{\perp\perp} = \{L : \text{every non–zero submodule of } L \text{ has a non–zero sub–submodule in } \mathscr{X}\}$.

In this situation, we shall call $\mathscr{X}^{\perp}$ and $\mathscr{X}^{\perp\perp}$ the $\mathscr{X}$*-void* and $\mathscr{X}$*-full* classes, respectively.

Definition 1.32. A module P is called *purely infinite* if $P \cong P \oplus P$.

Lemma 1.33. *Let* P *be a purely infinite module. If* $B \rightarrowtail P$, *then* $B^{(\mathbb{N})} \rightarrowtail P$.

PROOF. It is easy to see that $P^{(\mathbb{N})} \rightarrowtail P$; hence $B^{(\mathbb{N})} \rightarrowtail P$. □

Lemma 1.34. *Let* $E = B \oplus P$ *where* E *is injective and* P *is purely infinite. If* $B \rightarrowtail P$, *then* $E \cong P$.

PROOF. By the previous lemma $B^{(\mathbb{N})} \rightarrowtail P$; hence $P \geq \bigoplus_{i=1}^{\infty} B_i$ with $B_i \cong B$. Since B and P are injective

$$P = E(\bigoplus_{i=1}^{\infty} B_i) \oplus C = E(B_1) \oplus E(\bigoplus_{i=2}^{\infty} B_i) \oplus C$$

$$\cong B \oplus E(\bigoplus_{i=1}^{\infty} B_i) \oplus C = B \oplus P = E.$$ □

Now we have all the ingredients for the decomposition theorem.

Theorem 1.35. *Every injective module* E *has a decomposition, unique up to superspectivity,* $E = D \oplus P$, *where* D *is directly finite,* P *purely infinite, and* D *and* P *are orthogonal.*

PROOF. We first prove the existence of the decomposition. Consider the hereditary class $\mathcal{X} = \{X : X^{(\mathbb{N})} \rightarrowtail E\}$. Then $E = V \oplus L$ where V is $\mathcal{X}$-void and L is $\mathcal{X}$-full. By construction V and L are orthogonal. Also V is directly finite by Proposition 1.27.

We proceed now to show that L is purely infinite. By Zorn's Lemma, there exists a maximal direct sum $K = \bigoplus_{\alpha \in \Lambda} Y_\alpha$ in L, where each Y_α is isomorphic to an infinite copower $X_\alpha^{(\mathbb{N})}$. Hence

$$K \cong \bigoplus_{\alpha \in \Lambda} X_\alpha^{(\mathbb{N})} \cong (\bigoplus_{\alpha \in \Lambda} X_\alpha)^{(\mathbb{N})} = \bigoplus_{i=1}^{\infty} Z_i,$$

where $Z_i \cong \bigoplus_{\alpha \in \Lambda} X_\alpha$, for all $i \in \mathbb{N}$. Let

$$K_1 = \bigoplus_{n=1}^{\infty} Z_{2n-1} \text{ and } K_2 = \bigoplus_{n=1}^{\infty} Z_{2n}.$$

Then $K = K_1 \oplus K_2$ and $K_1 \cong K \cong K_2$. Since L is injective, $L = F \oplus N$ where $N = E(K)$. Maximality of K then implies F is directly finite. Also

$$N = E(K) = E(K_1) \oplus E(K_2) \cong N \oplus N,$$

hence N is purely infinite.

Another application of Zorn's Lemma yields a monomorphism $F \geq H \overset{f}{\rightarrowtail} N$ that cannot be extended. Since L is injective, $H \subset^{\oplus} F$ and $fH \subset^{\oplus} N$. Write $F = H \oplus H'$ and $N = fH \oplus N'$. Then

$$fH \oplus N' = N \cong N \oplus N = fH \oplus N' \oplus N.$$

Since fH, being isomorphic to H, is injective and directly finite, we get by Theorem 1.29 that $N' \cong N' \oplus N$. Since $N' \leq N$ and N is purely infinite, $N' \cong N$ by Lemma 1.34.

Now we claim that $H' = 0$. Suppose not. As H' is $\mathcal{X}$-full, H' contains a non–zero submodule W such that $W^{(\infty)} \rightarrowtail E$. Then $E \geq \bigoplus_{j=1}^{\infty} W_j$ such that $W_j \cong W$ for every $j \in \mathbb{N}$. Since V is $\mathcal{X}$-void, $V \cap (\bigoplus_{j=1}^{\infty} W_j) = 0$ by Lemma 1.31. Hence $\bigoplus_{j=1}^{\infty} W_j \rightarrowtail L$. Since F is directly finite, $N \cap (\bigoplus_{j=1}^{\infty} W_j) \neq 0$; otherwise $\bigoplus_{j=1}^{\infty} W_j \rightarrowtail F$, a contradiction. Applying Lemma 1.31, we get $t \in \mathbb{N}$ and $0 \neq T \leq N$ that

As $T \leq N \cong N'$, we get a non–zero monomorphism $H' \geq H'' \overset{g}{\rightarrowtail} N'$. But then

$$H \oplus H'' \xrightarrow{f \oplus g} fH \oplus N' = N$$

extends f, a contradiction. Hence $H' = 0$, and so $F = H$. Thus $F \rightarrowtail N$, and by Lemma 1.34, $L = F \oplus N \cong N$. We conclude that L is purely infinite.

Now we prove the uniqueness. We will establish this by proving that for any other decomposition $E = D \oplus P$ with the given properties, D is $\mathcal{X}$-void and P is $\mathcal{X}$-full. Since $P^{(\mathbb{N})} \rightarrowtail P \leq E$, we have $P \in \mathcal{X}$ and consequently P is $\mathcal{X}$-full. Now consider D, and let $0 \neq X \in \mathcal{X}$ such that $X \leq D$. Thus $E \geq \bigoplus_{i=1}^{\infty} X_i$ with $X_i \cong X$ for every $i \in N$. If $P \cap (\bigoplus_{i=1}^{\infty} X_i) = 0$, then $\bigoplus_{i=1}^{\infty} X_i \rightarrowtail D$, a contradiction since D is directly finite. On the other hand $P \cap (\bigoplus_{i=1}^{\infty} X_i) \neq 0$ yields by Lemma 1.31 that there exists $0 \neq P' \leq P$ and $t \in \mathbb{N}$ such that $P' \rightarrowtail X_t$. Then

$$P \geq P' \rightarrowtail X_t \cong X \leq D$$

which is again a contradiction since P and D have no non–zero isomorphic submodules. Therefore $X = 0$ and hence D is $\mathcal{X}$-void. Hence the decomposition $M = L \oplus V$ is unique up to superspectivity. □

COMMENTS

Baer [40] initiated the study of abelian groups which are summands whenever they are subgroups. These are precisely the divisible abelian groups, that is abelian groups G with $nG = G$ for every $n \in \mathbb{N}$. Modules which are summands of every containing module were studied by a number of authors (initially under several different names, e.g. algebraicly compact, M_u–modules, ... etc). Eckmann and Schopf [53] introduced the terminology "injective"; they also proved the existence of the injective hull.

Johnson and Wong [61] defined the notion of a quasi–injective module. They proved that a module is quasi–injective if and only if it is closed under all endomorphisms of its injective hull, and hence any module M has a quasi–injective hull (End E(M))M. Our proof of Lemma 1.13 is essentially the one given by Johnson and Wong for quasi–injective modules.

(Quasi)–injective modules were studied extensively. A number of generalizations were defined and studied by many authors, e.g. pseudo–injective modules (Singh [67]), Ker–injective modules (Birkenmeier [78]), π–injective modules (Goel and Jain [78]). Another generalization of (quasi–)injectivity will be discussed in Chapter 2.

Rings for which a certain class of modules satisfies some "generalized injectivity property" were studied by many authors: Ahsan, Birkenmeier, Boyle, Byrd, Faith, Goel, Jain, Koehler, Michler, Mohamed, Müller, Osofsky, Saleh, Singh, Symmonds, Villamayor, and others.

In this chapter we concentrated our attention on one generalization of injectivity, namely A–injectivity (Definition 1.1), as it is indispensible for our study of (quasi–)continuous modules in later chapters. This notion was studied by Azumaya [P], Azumaya et al. [75], Sandomierski [64] and de Robert [69].

The proofs of 1.2 to 1.6 are essentially the same as for the analogous results for injective modules. Theorems 1.7 and 1.11 are results of Azumaya et al. [75], modified by some ideas in Müller and Rizvi [84]. The rest of the results in this section are modified versions of those obtained by Müller and Rizvi [84].

Theorem 1.21 was proved by Warfield [69a] for injective modules, and the proof was generalized to quasi–injective modules by Fuchs [69]; the proof presented here is slightly different. Proposition 1.23 is also due to Fuchs [72]. Suzuki [68] proved Proposition 1.28 by utilizing the properties of the endomorphism rings of injective modules; we give a different and a more direct proof. Using Suzuki's result, Birkenmeier [76] proved the conclusion of Theorem 1.29 for quasi–injective modules; this theorem will be generalized to continuous modules in Chapter 3.

Goodearl and Boyle [76] proved Theorem 1.35 for non–singular injective modules, where the decomposition is unique. The existence of the decomposition for arbitrary injective modules is contained in Goodearl [79], where the full theorem is obtained from its non–singular special case via some functorial techniques. A direct proof is given in Müller and Rizvi [83]; the proof included here is considerably shorter and depends on the class decomposition developed in this chapter.

CHAPTER 2

QUASI–CONTINUOUS MODULES

In this chapter we discuss generalizations of the notion of continuous rings studied by von Neumann [36] and Utumi [65] to modules. Such modules are also generalizations of (quasi–) injective modules.

1. BASIC PROPERTIES

Proposition 2.1. *Any (quasi–)injective module* M *satisfies the following two conditions:*

(C_1) *Every submodule of* M *is essential in a summand of* M*;*

(C_2) *If a submodule* A *of* M *is isomorphic to a summand of* M, *then* A *is a summand of* M.

PROOF. Let $N \leq M$ and write $E(M) = E_1 \oplus E_2$ where $E_1 = E(N)$. The quasi–injectivity of M implies, by Corollary 1.14, that $M = M \cap E_1 \oplus M \cap E_2$; and it is clear that $N \leq^e M \cap E_1$. Hence (C_1) holds.

Let $M' \overset{f}{\rightarrowtail} M$ be a monomorphism with $M' \subset^{\oplus} M$. Since M is M–injective, M′ is M–injective by Proposition 1.6. Then f splits by Lemma 1.2; thus (C_2) holds.

□

Proposition 2.2. *If a module* M *has* (C_2), *then it satisfies the following condition:*

(C_3) *If* M_1 *and* M_2 *are summands of* M *such that* $M_1 \cap M_2 = 0$, *then* $M_1 \oplus M_2$ *is a summand of* M.

PROOF. Write $M = M_1 \oplus M_1^*$ and let π denote the projection $M_1 \oplus M_1^* \twoheadrightarrow M_1^*$. Then $M_1 \oplus M_2 = M_1 \oplus \pi M_2$. Since $\pi|_{M_2}$ is a monomorphism, we get $\pi M_2 \subset^{\oplus} M$ by (C_2). As $\pi M_2 \leq M_1^*$, $M_1 \oplus \pi M_2 \subset^{\oplus} M$. □

Definition 2.3. A module M is called *continuous* if it has (C_1) and (C_2); M is called *quasi–continuous* if it has (C_1) and (C_3).

We have just seen that the following implications hold:

Injective ⇒ quasi–injective ⇒ continuous ⇒ quasi–continuous ⇒ (C_1).

To illustrate this hierarchy of concepts, and at the same time demonstrate that they are all distinct, we list now, without proof, all abelian groups with these properties. (Proofs are easily obtained from the following development, and from Kamal and Müller [88a]). We also display how regular rings, as right modules over themselves, fit into the scheme.

Abelian Groups	Concept	Regular Rings
$\oplus\{\mathbb{Q}, C_p^\infty \text{ (all p)}\}$	Injective	Right self injective
	Quasi-injective	
above, and $\oplus\{C_p^{n(p)} \text{ (all p)}\}$	Continuous	Right continuous
above, and X $\oplus$ torsion injective ($X \leq \mathbb{Q}$)	Quasi-continuous	
above, and $X^n \oplus$ injective, and $\oplus\{C_p^{n(p)}, C_p^{n(p)+1} \text{ or } C_p^\infty \text{ (all p)}\}$	(C_1)	

For a discussion, and a certain amount of classification of self-injective and continuous regular rings, cf. Goodearl [79]. Simple examples of a self-injective regular ring, and a continuous regular ring which is not self-injective, are $\Pi\, F_\alpha$ (where $\{F_\alpha\}$ is an infinite family of fields), and its subring $\Pi(F_\alpha, P_\alpha)$ (where the P_α are proper subfields of the F_α) consisting of all the sequences with almost all entries in P_α, cf. Utumi [60].

A submodule of a module M is *closed* if it has no proper essential extensions in M. A submodule X of M is a *complement* if it is maximal with respect to $X \cap Y = 0$, for some submodule Y. Closed and complement submodules are the same; they exist in abundance, by Zorn's Lemma.

The following two results are obvious.

Proposition 2.4. *A module* M *has* (C_1) *if and only if every closed submodule of* M *is a summand.* □

Proposition 2.5. *An indecomposable module* M *has* (C_1) *if and only if* M *is uniform. Any uniform module is quasi–continuous.* □

Lemma 2.6. *Let* A *be a submodule of an arbitrary module* M. *If* A *is closed in a summand of* M, *then* A *is closed in* M.

PROOF. Let $M = M_1 \oplus M_2$ with A closed in M_1. Let π denote the projection $M_1 \oplus M_2 \longrightarrow\!\!> M_1$. Assume that $A \leq^e B$ for some $B \leq M$. Then it is easy to see that $A = \pi A \leq^e \pi B \leq M_1$. Since A is closed in M_1, $\pi B = A \leq B$, and so, $(1-\pi)B \leq B$. Since $(1 - \pi)B \cap A = 0$ and $A \leq^e B$, $(1 - \pi)B = 0$ and hence $B = \pi B \leq M_1$. Then $A = B$ since A is closed in M_1. □

Proposition 2.7. *The conditions* (C_i) $(i = 1,2,3)$ *are inherited by summands. In particular, any summand of a (quasi–)continuous module is (quasi–)continuous.*

PROOF. Follows from the definitions and Lemma 2.6. □

We end this section by giving characterizations of quasi–continuous modules, in terms of their injective hulls, and their complement submodules.

Theorem 2.8. *The following are equivalent for a module* M:

(1) M *is quasi–continuous;*

(2) $M = X \oplus Y$ *for any two submodules* X *and* Y *which are complements of each other;*

(3) $fM \leq M$ *for every idempotent* $f \in \text{End } E(M)$;

(4) $E(M) = \bigoplus_{i \in I} E_i$ *implies* $M = \bigoplus_{i \in I} M \cap E_i$.

PROOF. (1) ⇒ (2): Now $X, Y \subset^{\oplus} M$ by Proposition 2.4, and hence $X \oplus Y \subset^{\oplus} M$ by (C_3). Since $X \oplus Y \leq^e M$, $M = X \oplus Y$.

(2) ⇒ (3): Let $A_1 = M \cap f\,E(M)$ and $A_2 = M \cap (1-f)\,E(M)$. Let B_1 be a complement of A_2 that contains A_1, and let B_2 be a complement of B_1 that contains A_2. Then $M = B_1 \oplus B_2$. Let π be the projection $B_1 \oplus B_2 \longrightarrow\!\!> B_1$. We claim that $M \cap (f - \pi)\,M = 0$. Let $x, y \in M$ be such that $(f - \pi)(x) = y$. Then

$f(x) = y + \pi(x) \in M$, and hence $f(x) \in A_1$. Thus $(1 - f)x \in M$ and so $(1 - f)x \in A_2$. Therefore $\pi(x) = f(x)$, and consequently $y = 0$. This proves our claim. Since $M \leq^e E(M)$, $(f - \pi)M = 0$, and so $fM = \pi M \leq M$.

(3) ⇒ (4): It is clear that $\bigoplus_{i \in I} M \cap E_i \leq M$. Let m be an arbitrary element in M. Then $m \in \bigoplus_{i \in F} E_i$ for a finite subset $F \subseteq I$. Write $E(M) = \bigoplus_{i \in F} E_i \oplus E^*$. Then there exist orthogonal idempotents $f_i \in \text{End } E(M)$ $(i \in F)$ such that $E_i = f_i E(M)$. Since $f_i M \leq M$ by assumption,

$$m = (\sum_{i \in F} f_i)(m) = \sum_{i \in F} f_i(m) \in \bigoplus_{i \in F} M \cap E_i.$$

Thus $M \leq \bigoplus_{i \in I} M \cap E_i$ and therefore $M = \bigoplus_{i \in I} M \cap E_i$.

(4) ⇒ (1): Let $A \leq M$. Write $E(M) = E(A) \oplus E^*$. Then $M = M \cap E(A) \oplus M \cap E^*$ with $A \leq^e M \cap E(A)$. Thus M has (C_1). Let $M_1, M_2 \subset^{\oplus} M$ with $M_1 \cap M_2 = 0$. Write $E(M) = E_1 \oplus E_2 \oplus E'$ where $E_i = E(M_i)$, $i = 1,2$. Then

$$M = M \cap E_1 \oplus M \cap E_2 \oplus M \cap E'.$$

Since $M_i \subset^{\oplus} M$ and $M_i \leq^e M \cap E_i$, $M_i = M \cap E_i$ $(i = 1,2)$. Therefore M has (C_3).

□

2. DIRECT SUMS OF QUASI–CONTINUOUS MODULES

A summand of a quasi–continuous module is quasi–continuous (Proposition 2.7). However a direct sum of quasi–continuous modules need not be quasi–continuous. This is illustrated by the following:

__Example 2.9__. Let $R = \begin{bmatrix} F & F \\ 0 & F \end{bmatrix}$ where F is any field. Let $A = \begin{bmatrix} F & F \\ 0 & 0 \end{bmatrix}$ and $B = \begin{bmatrix} 0 & 0 \\ 0 & F \end{bmatrix}$. It is clear that A and B are quasi–continuous as R–modules (in fact A is injective and B is simple). However $R = A \oplus B$ is not quasi–continuous. (It is easy to check that R_R satisfies (C_1) but does not satisfy (C_3).) □

The following proposition gives a necessary condition for $M_1 \oplus M_2$ to be quasi–continuous.

__Proposition 2.10__. *If $M_1 \oplus M_2$ is quasi–continuous, then M_1 and M_2 are relatively injective.*

PROOF. We show M_2 is M_1–injective. Write $M = M_1 \oplus M_2$. Let $X \leq M_1$ and $\varphi : X \longrightarrow M_2$ be a homomorphism and let $B = \{x - \varphi(x) : x \in X\}$. It is obvious that $B \cap M_2 = 0$. Let M_1^* be a complement of M_2 that contains B. Then $M = M_1^* \oplus M_2$ by Theorem 2.8. Let π denote the projection $M_1^* \oplus M_2 \longrightarrow\!\!\!> M_2$. For all $x \in X$ we have $0 = \pi\,(x - \varphi(x)) = \pi(x) - \pi(\varphi(x)) = \pi(x) - \varphi(x)$. Hence $\pi|_{M_1}$ extends φ. □

<u>Corollary 2.11</u>. *If* $M_1 \oplus M_2$ *is quasi–continuous and* $M_1 \rightarrowtail M_2$, *then* M_1 *is quasi–injective.* M *is quasi–injective if and only if* $M \oplus M$ *is quasi–continuous.* □

<u>Corollary 2.12</u>. *A purely infinite module* M *is quasi–injective if and only if* M *is quasi–continuous.* □

Now we consider a direct sum $M = \bigoplus_{\alpha \in \Lambda} M_\alpha$ of quasi–continuous modules M_α. A necessary condition for M to be quasi–continuous is that $M(\Lambda - \alpha)$ is M_α–injective for every $\alpha \in \Lambda$. In the following we show that this condition is also sufficient.

<u>Theorem 2.13</u>. *Let* $\{M_\alpha : \alpha \in \Lambda\}$ *be a family of quasi–continuous modules. Then the following are equivalent:*

(1) $M = \bigoplus_{\alpha \in \Lambda} M_\alpha$ *is quasi–continuous;*

(2) $M(\Lambda - \alpha)$ *is* M_α*–injective for every* $\alpha \in \Lambda$;

(3) M_α is M_β*–injective for all* $\alpha \neq \beta \in \Lambda$ *and* (A_2) *holds.*

PROOF. (2) ⇔ (3) follows by Proposition 1.9, and (1) ⇒ (2) by Proposition 2.10. It remains to show (2) ⇒ (1). In view of Theorem 2.8, we have to show that $eM \leq M$ for every idempotent $e \in \text{End } E(M)$. Since

$$eM = e\Big(\bigoplus_{\alpha \in \Lambda} M_\alpha\Big) = \sum_{\alpha \in \Lambda} e\,M_\alpha,$$

we need only show that $eM_\alpha \leq M$ for every $\alpha \in \Lambda$.

Consider a fixed $\alpha \in \Lambda$. Since M_α is M_β–injective for all $\alpha \neq \beta \in \Lambda$, M_α is $M(\Lambda - \alpha)$–injective by Proposition 1.5. Write $N_1 = M_\alpha$ and $N_2 = M(\Lambda - \alpha)$. Then

N_1 and N_2 are relatively injective and N_1 is quasi–continuous. Let E, E_1 and E_2 denote the injective hulls of M, N_1 and N_2, respectively. Then $E = E_1 \oplus E_2$ and $e = \begin{bmatrix} e_{11} & e_{12} \\ e_{21} & e_{22} \end{bmatrix}$ where $e_{ij} : E_j \longrightarrow E_i$. Since N_2 is N_1–injective, $e_{21}N_1 \leq N_2$ by Lemma 1.13; consequently

$$eN_1 = e_{11}N_1 + e_{21}N_1 \leq e_{11}N_1 + N_2.$$

Thus it is enough to show that $e_{11}N_1 \leq M$.

Since $e = e^2$, $e_{11} = e_{11}^2 + e_{12}\, e_{21}$. Write $a = e_{11}$ and $b = 1 - e_{11}$. Then

$$ab = ba = a - a^2 = e_{12}e_{21} \in \text{End } E_1.$$

Let K = Ker ab; then it is obvious that $aK \cap bK = 0$ and $aK \leq \text{Ker } b \leq \text{Ker } ab = K$. Hence $K = aK \oplus bK$. Since E_1 is injective, there exist orthogonal idempotents f and g in End E_1 such that $E_1 = fE_1 \oplus gE_1$, $aK \leq fE_1$ and $bK \leq gE_1$. Then

$$fK = f(aK \oplus bK) = faK = aK.$$

Therefore $K \cap fE_1 \leq fK = aK \leq K \cap fE_1$; and consequently

$$K \cap fE_1 = aK \leq \text{Ker } b.$$

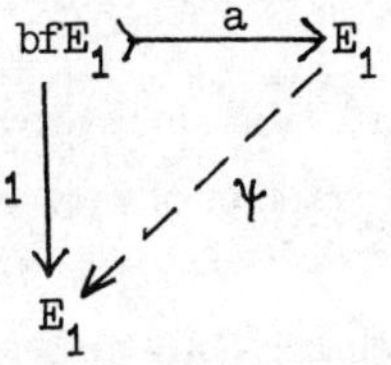

Hence $a|_{bfE_1}$ is a monomorphism. Since E_1 is injective, there exists $\psi \in \text{End } E_1$ such that $bf = \psi abf$.

$$\begin{array}{ccc} bfE_1 & \xrightarrow{a} & E_1 \\ {\scriptstyle 1}\downarrow & \swarrow {\scriptstyle \psi} & \\ E_1 & & \end{array}$$

Since N_1 is quasi–continuous and N_1 and N_2 are relatively injective, we get by Theorem 2.8 and Lemma 1.13

$$bfN_1 = \psi abfN_1 \leq \psi abN_1 = \psi e_{12}e_{21}N_1 \leq \psi e_{12}N_2 \leq N_1.$$

Similarly one can prove that $agN_1 \leq N_1$. Then

$$aN_1 = a(f + g)N_1 = afN_1 + agN_1 = (1 - b)fN_1 + agN_1 \leq N_1.$$

Hence $e_{11}N_1 = aN_1 \leq N_1$. □

<u>Corollary 2.14</u>. $\bigoplus_{i=1}^{n} M_i$ *is quasi–continuous if and only if each* M_i *is quasi–continuous and* M_j*–injective for all* $j \neq i$. □

3. DECOMPOSITIONS OF QUASI–CONTINUOUS MODULES

In general, a quasi–continuous module need not be a direct sum of indecomposable submodules. On the other hand, a quasi–continuous module which is a direct sum of indecomposable submodules, behaves in many ways as if these submodules had local endomorphism rings, though this need not be the case. Here we characterize those quasi–continuous modules which are direct sums of indecomposable modules. We also discuss briefly some fundamental properties of arbitrary modules which are direct sums of indecomposable submodules.

<u>Definition 2.15</u>. A family $\{X_\lambda : \lambda \in \Lambda\}$ of submodules of a module M is called a *local summand* of M, if $\sum_{\lambda\in\Lambda} X_\lambda$ is direct and $\sum_{\lambda\in F} X_\lambda$ is a summand of M for every finite subset $F \subseteq \Lambda$. (When there is no danger of confusion we simply say that $\sum_{\lambda\in\Lambda} X_\lambda$ is a local summand of M). If even $\sum_{\lambda\in\Lambda} X_\lambda$ is a summand of M, we say that *the local summand is a summand.*

We shall prove that if every local summand of M is a summand, then M is a direct sum of indecomposable submodules (Theorem 2.17). First we need a lemma:

<u>Lemma 2.16</u>. *Let* M *be an arbitrary module. Every local summand of* M *is a summand if and only if the union of any chain of summands of* M *is a summand.*

PROOF. "If" : Let $\mathscr{X} = \{X_\lambda : \lambda \in \Lambda\}$ be a local summand of M. Let $\mathscr{L}$ be the family of all subsets Ω of Λ such that $\sum_{\lambda\in\Omega\cup F} X_\lambda \subset^{\oplus} M$ for every finite subset $F \subseteq \Lambda$. Consider a chain Ω_α in $\mathscr{L}$, and let $\tilde{\Omega} = \bigcup_\alpha \Omega_\alpha$. Then for any finite subset $F \subseteq \Lambda$,

$$\sum_{\lambda\in\tilde{\Omega}\cup F} X_\lambda = \bigcup_\alpha \Big(\sum_{\lambda\in\Omega_\alpha\cup F} X_\lambda \Big)$$

is a summand of M by assumption. Thus $\tilde{\Omega} \in \mathscr{L}$, and hence is an upper bound of the chain.

Then Zorn's Lemma applies to $\mathscr{L}$ and we get a maximal element Ω. Suppose $\Omega \neq \Lambda$. Let $\gamma \in \Omega - \Lambda$ and let $\Omega^+ = \Omega \cup \{\gamma\}$. For any finite subset $F \subseteq \Lambda$,

$$\sum_{\lambda\in\Omega^+\cup F} X_\lambda = \sum_{\lambda\in\Omega\cup(\{\gamma\}\cup F)} X_\lambda \subset^{\oplus} M.$$

It then follows that $\Omega^+ \in \mathscr{L}$, which contradicts the maximality of Ω. Hence $\Omega = \Lambda$, and $\sum_{\lambda\in\Lambda} X_\lambda \subset^{\oplus} M$.

"Only if" : Let $\{M_i : i \in I\}$ be a chain of summands of M. Consider the collection $\mathcal{F}$ of all local summands $\mathcal{X} = \{X_\lambda : \lambda \in \Lambda\}$ with the additional property $\sum_{\lambda \in \Lambda} X_\lambda = \bigcup_{i \in J} M_i$ for some subset $J = J(\mathcal{X})$ of I. We order $\mathcal{F}$ by inclusion. Consider a chain $\mathcal{X}_\alpha = \{X_\lambda : \lambda \in \Lambda_\alpha\}$ in $\mathcal{F}$ and let $\tilde{\mathcal{X}} = \bigcup \mathcal{X}_\alpha = \{X_\lambda : \lambda \in \bigcup \Lambda_\alpha\}$. It is clear that $\tilde{\mathcal{X}}$ is again a local summand. Now

$$\sum_{\lambda \in \cup \Lambda_\alpha} X_\lambda = \bigcup_\alpha \Big(\sum_{\lambda \in \Lambda_\alpha} X_\lambda \Big) = \bigcup_\alpha \Big(\bigcup_{i \in J(\mathcal{X}_\alpha)} M_i \Big) = \bigcup_{i \in \cup_\alpha J(\mathcal{X}_\alpha)} M_i .$$

Thus $\tilde{\mathcal{X}} \in \mathcal{F}$, and hence is an upper bound of the chain.

If follows by Zorn's Lemma that $\mathcal{F}$ has a maximal element $\mathcal{X}$. Let $A = \sum_{\lambda \in \Lambda} X_\lambda = \bigcup_{i \in J(\mathcal{X})} M_i$. We claim that $A = \bigcup_{i \in I} M_i$. Suppose not. Then there exists $k \in I$ such that $M_k \nleq A$. Hence $M_k \nleq M_i$ and therefore $M_i < M_k$ for all $i \in J(\mathcal{X})$. Since $A \subset^{\oplus} M$ by hypothesis, $M_k = A \oplus B$, for some $0 \neq B \leq M$. Then it is clear that $\mathcal{X} \cup \{B\} \in \mathcal{F}$, which contradicts the maximality of $\mathcal{X}$. □

Theorem 2.17. *If every local summand of a module* M *is a summand, then* M *is a direct sum of indecomposable modules.*

PROOF. By Zorn's Lemma, M contains a maximal local summand $\mathcal{X} = \{X_\lambda : \lambda \in \Lambda\}$ where each X_λ is indecomposable. Let $X = \sum_{\lambda \in \Lambda} X_\lambda$. Then $X \subset^{\oplus} M$ by assumption; write $M = X \oplus Y$. We claim that $Y = 0$. To the contrary, assume that $Y \neq 0$ and consider a non-zero element $y \in Y$. By Lemma 2.16, there exists a summand A of Y maximal such that $y \notin A$. Then $Y = A \oplus B$ with $B \neq 0$. Now maximality of $\mathcal{X}$ forces B to be decomposable; so $B = B_1 \oplus B_2$ with $B_i \neq 0$ $(i = 1,2)$. Again maximality of A implies $y \in A \oplus B_i$ $(i = 1,2)$. But then $y \in A$, a contradiction. Thus $Y = 0$ and $M = X = \bigoplus_{\lambda \in \Lambda} X_\lambda$. □

Proposition 2.18. *Let* M *be an* R*–module with* (C_1). *If* R *satisfies the ascending chain condition on right ideals of the form* m^o, $m \in M$, *then every local summand of* M *is a summand.*

PROOF. Let $\mathcal{X} = \{X_\lambda : \lambda \in \Lambda\}$ be a local summand of M, and $X = \sum_{\lambda \in \Lambda} X_\lambda$. If X^* is a closure of X in M, then $X^* \subset^{\oplus} M$ and hence X^* has (C_1). So there is no loss of generality if we assume that $X \leq^e M$; in that case we have to show that $X = M$.

Suppose that $X \neq M$ and select $m \in M - X$ such that m^o is maximal. Since $X \leq^e M$, there exists $r \in R$ such that $0 \neq mr \in X$. Now $mr \in \bigoplus_{\lambda \in F} X_\lambda$ for some finite subset $F \subseteq \Lambda$. By assumption, $M = \bigoplus_{\lambda \in F} X_\lambda \oplus Y$ for some $Y \leq M$. Then $m = x + y$, $x \in \bigoplus_{\lambda \in F} X_\lambda$ and $y \in Y$. Clearly $y \notin X$ and $m^o \leq y^o$; hence $m^o = y^o$ by maximality of m^o. Since

$$yr = mr - xr \in (\bigoplus_{\lambda \in F} X_\lambda) \cap Y = 0,$$

$mr = 0$, a contradiction. Hence $X = M$. □

A ring R is right noetherian if and only if every injective R–module is a direct sum of uniform modules (Matlis [58], Papp [59]). The following generalizes this result to modules with (C_1).

Theorem 2.19. *If* R *is rihgt noetherian, then every* R*–module with* (C_1) *is a direct sum of uniform modules.*

PROOF. The result follows from 2.5, 2.17 and 2.18. □

If a module M has a decomposition $M = \bigoplus_{i \in I} M_i$ where each M_i has a local endomorphism ring, then the Krull–Schmidt–Azumaya Theorem asserts that this decomposition is unique up to isomorphism (i.e. if $M = \bigoplus_{j \in J} N_j$ is another decomposition of M with indecomposable N_j, then there exist an automorphism φ of M and a bijection $\pi : I \longrightarrow J$ such that $N_{\pi(i)} = \varphi M_i$). The conclusion of this theorem remains true in some cases where the M_i are indecomposable but no longer have local endomorphism rings. One example is the class of modules with a decomposition that complements summands. (c.f. Anderson and Fuller [73], Ch. 3, §12.) For the reader's convenience we include the definitions and results which will be used in this book.

(1) A submodule N of M is called a *maximal summand* in M, if $M = N \oplus N'$ with N' indecomposable.

(2) A decomposition $M = \bigoplus_{i \in I} M_i$ is said to *complement* (*maximal*) *summands* if for every (maximal) summand A of M there exists a subset $J \subseteq I$ such that $M = A \oplus (\bigoplus_{i \in J} M_i)$.

(3) Let $M = \bigoplus_{i \in I} M_i$, with each M_i indecomposable, be a decomposition that complements (maximal) summands. Then:

(a) This decomposition satisfies the conclusion of the Krull–Schmidt–Azumaya Theorem;

(b) Any other decomposition of M into indecomposables complements (maximal) summands.

(4) Let $M = \bigoplus_{i \in I} M_i$ be a decomposition that complements summands. Then:

(a) M_i is indecomposable for every $i \in I$;

(b) Every summand of M has a decomposition that complements summands;

(c) If $M = \bigoplus_{\alpha \in \Lambda} A_\alpha$, then I is the disjoint union of subsets I_α such that $A_\alpha \cong \bigoplus_{i \in I_\alpha} M_i$ $(\alpha \in \Lambda)$.

Definition 2.20. A module M is said to have the (*finite*) *extending property* if for any (finite) index set I, and any direct sum $\bigoplus_{i \in I} A_i$ of submodules A_i of M, there exists a family $\{M_i : i \in I\}$ of submodules of M such that $A_i \leq^e M_i$ and $\bigoplus_{i \in I} M_i$ is a summand of M.

For $n \in \mathbb{N}$, the n–*extending property* for M will mean that M has the extending property relative to index sets of cardinality n.

The proof of the following lemma is straightforward.

Lemma 2.21. *Let* M *be an arbitrary module. Then:*

(1) M *has the* 1–*extending property if and only if* M *has* (C_1);

(2) M *has the finite extending property, if and only if* M *has the* 2–*extending property, if and only if* M *is quasi–continuous;*

(3) M *has the extending property if and only if* M *has the finite extending property and every local summand of* M *is a summand.* □

Theorem 2.22. *Let* M *be a quasi–continuous module. Then the following are equivalent:*

(1) M *is a direct sum of indecomposable (uniform) modules*;

(2) M *has a decomposition that complements summands*;

(3) *Every local summand of* M *is a summand*;

(4) M *has the extending property.*

PROOF. (1) ⇒ (2): Let $M = \bigoplus_{i\in I} M_i$ be a decomposition of M into indecomposable (hence uniform) submodules M_i; we show that this decomposition complements summands. Let N be any summand of M. By Zorn's Lemma, there exists a subset $K \subseteq I$ maximal such that $N \cap M(K) = 0$. Since M is quasi-continuous, $N \oplus M(K) \subset^{\oplus} M$. Next we prove that $N \oplus M(K) \leq^e M$; it then follows that $M = N \oplus M(K)$. Let $i \in I$; then maximality of K implies $X_i = (N \oplus M(K)) \cap M_i \neq 0$. Since M_i is uniform, $X_i \leq^e M_i$, and consequently

$$\bigoplus_{i\in I} X_i \leq^e \bigoplus_{i\in I} M_i = M.$$

As $N \oplus M(K) \geq \bigoplus_{i\in I} X_i$, $N \oplus M(K) \leq^e M$. This proves our claim, and hence the decomposition complements summands.

(2) ⇒ (3): Let $M = \bigoplus_{\alpha\in\Lambda} M_\alpha$ be a decomposition that complements summands. Let $A = \bigoplus_{j\in J} A_j$ be a local summand of M. Since M is quasi-continuous, $A \leq^e A^* \subset^{\oplus} M$; and if we write $M = A^* \oplus B$, we get that $\bigoplus_{j\in J} A_j \oplus B$ is a local summand of M and essential in M. So there is no loss of generality if we assume $A \leq^e M$; in that case we have to show $A = M$.

Suppose $A \neq M$. Inductively we construct a sequence of elements x_n such that $x_n \notin A$, $x_n \in M_{\alpha_n}$, the α_n are distinct, and

$$x_1^o < x_2^o < \ldots < x_n^o < \ldots .$$

Once this is done, we get a contradiction to (A_3), and hence to (A_2), which is valid by Theorem 2.13. (Recall the conditions (A_i) from Section 1 of Chapter 1).

Assume that $x_1, x_2, \ldots, x_n$ have been constructed. Since $A \leq^e M$, there exist $r_t \in R$ such that $0 \neq x_t r_t \in A$, $t = 1,2,\ldots,n$. There exists a finite subset $F \subseteq J$ such that $x_t r_t \in \bigoplus_{j\in F} A_j$ for all $t = 1,2,\ldots,n$. Since $A = \bigoplus_{j\in J} A_j$ is a local summand of M and the decomposition $M = \bigoplus_{\alpha\in\Lambda} M_\alpha$ complements summands, there exists a subset $K \subseteq \Lambda$

such that

$$M = \bigoplus_{j\in F} A_j \oplus M(K).$$

Write $x_n = a + \Sigma\, y_\alpha$ where $a \in \bigoplus_{j\in F} A_j$ and $\Sigma\, y_\alpha \in M(K)$. It is clear that $x_n^o \leq y_\alpha^o$ for every α. Now

$$(\Sigma\, y_\alpha)\, r_n = x_n r_n - a r_n \in (\bigoplus_{j\in F} A_j) \cap M(K) = 0,$$

and hence $x_n^o < y_\alpha^o$. Since $x_n \notin A$, there exists $\beta \in K$ such that $y_\beta \notin A$. Take $x_{n+1} = y_\beta$; then it is clear that x_{n+1} satisfies the required conditions.

(3) ⇒ (1) follows by Theorem 2.17, and (3) ⇔ (4) is a consequence of Lemma 2.21. □

We now state the definition of a concept which originated from the T–nilpotency that occurs in the study of perfect rings (Bass [60]). It is equivalent, under suitable assumptions, to a number of other interesting properties (see Theorem 2.25). Though it looks technical, it is usually the one condition that can be explicitly verified.

Definition 2.23. A family of modules $\{M_\alpha : \alpha\in\Lambda\}$ is called *locally–semi–transfinitely–nilpotent* (lsTn) if for any subfamily M_{α_i} $(i\in\mathbb{N})$ with distinct α_i and any family of non–isomorphisms $f_i : M_{\alpha_i} \longrightarrow M_{\alpha_{i+1}}$, and for every $x \in M_{\alpha_1}$, there exists $n\in\mathbb{N}$ (depending on x) such that $f_n \dots f_2 f_1(x) = 0$.

Proposition 2.24. *Let* $\{M_\alpha : \alpha\in\Lambda\}$ *be a family of uniform modules such that* M_α *is* M_β*–injective for all* $\beta \neq \alpha \in \Lambda$. *Then* (A_3) *is equivalent to* lsTn. *In particular both hold if* $\bigoplus_{\alpha\in\Lambda} M_\alpha$ *is quasi–continuous.*

PROOF. Let (A_3) be given and consider non–isomorphisms $f_n : M_{\alpha_n} \longrightarrow M_{\alpha_{n+1}}$ $(n \in \mathbb{N})$, with distinct α_n; and let $x \in M_{\alpha_1}$. Set $x_1 = x$ and $x_{n+1} = f_n \dots f_2 f_1(x)$. Then obviously $x_n^o \leq x_{n+1}^o$ and therefore $x_m^o = x_{m+1}^o$ holds for some $m \in \mathbb{N}$ by (A_3); consequently $f_m|_{x_m R} : x_m R \longrightarrow x_{m+1} R$ is an isomorphism. Assume that $x_m \neq 0$. Then since M_{α_m} is uniform, $f_m : M_{\alpha_m} \longrightarrow M_{\alpha_{m+1}}$ is a monomorphism. It then follows by Lemma 1.2 that f_m is an isomorphism, which is a contradiction. We conclude that $f_{m-1} \cdots f_2 f_1(x) = x_m = 0$.

Conversely, assume that $\{M_\alpha\}$ has lsTn, and consider elements $x_n \in M_{\alpha_n}$ ($n \in \mathbb{N}$) where the α_n are distinct and the sequence x_n^o is increasing. If the sequence x_n^o does not become stationary, then passing to a subsequence we may assume $x_n^o < x_{n+1}^o$ for all $n \in \mathbb{N}$. Then the natural maps $x_n R \longrightarrow x_{n+1} R$ are not monomorphisms, and hence their extensions by relative injectivity, $f_n : M_{\alpha_n} \longrightarrow M_{\alpha_{n+1}}$, are non–isomorphisms. By lsTn, $x_m = f_{m-1} \cdots f_2 f_1(x_1) = 0$ holds for some $m \in \mathbb{N}$. We conclude $x_m^o = R$, in contradiction to $x_m^o < x_{m+1}^o$.

The last statement follows from what we have already proved and Theorem 2.13. □

In this section we have discussed a number of properties for a module M which has a decomposition $M = \bigoplus_{\alpha \in \Lambda} M_\alpha$ into indecomposable submodules M_α. The following theorem, which we mention without proof, asserts that these properties are equivalent to each other, and to some other concepts like the exchange property, in case the M_α have local endomorphism rings. The equivalence of (1) to (4) is due to Harada [83a], and of (4) to (6) to Zimmermann–Huisgen and Zimmermann [84].

<u>**Theorem 2.25**</u>. *The following are equivalent for a module* M *with a decomposition* $M = \bigoplus_{\alpha \in \Lambda} M_\alpha$ *where each* M_α *has a local endomorphism ring*:

(1) *The radical factor ring* S/J(S) *of the endomorphism ring* S *of* M *is* (von Neumann) *regular, and idempotents lift* modulo J(S);

(2) *Every local summand of* M *is a summand*;

(3) *The decomposition complements summands*;

(4) $\{M_\alpha\}$ *is locally semi–T–nilpotent*;

(5) M *has the finite exchange property*;

(6) M *has the exchange property.* □

The implication (3) ⇒ (4) in this theorem holds without the assumption on the endomorphism rings of the M_α. This was proved by Kasch and Zöllner, and is apparently not published; we include a proof for the reader's convenience.

<u>**Theorem 2.26**</u>. *Any decomposition* $M = \bigoplus_{\alpha\in\Lambda} M_\alpha$ *which complements summands is locally semi–T–nilpotent.*

PROOF. It suffices to derive the conclusion of lsTn for non–isomorphisms $f_i : M_i \longrightarrow M_{i+1}$ ($i\in\mathbb{N}$). We may also assume that $M = \bigoplus_{i\in\mathbb{N}} M_i$. Let $M_i^* = (1-f_i)M_i$. The following properties are easy to obtain, and the verifications are left to the reader:

(a) $M_n^* \cap M_i = 0$ for $i \neq n$, and $M_n^* \cap M_n = 0$ if and only if f_n is a monomorphism;

(b) $\sum_{i\in\mathbb{N}} M_i^*$ is direct;

(c) $\bigoplus_{i=n}^{m} M_i^* \oplus M_{m+1} = \bigoplus_{i=n}^{m+1} M_i$.

For any subset $K \subseteq \mathbb{N}$, we write $M(K) = \bigoplus_{i\in K} M_i$ and $M^*(K) = \bigoplus_{i\in K} M_i^*$. We first show that lsTn holds if $M = M^*(\mathbb{N})$. Indeed given $a\in M_1$,

$$a = (1-f_1)m_1 + (1-f_2)m_2 + \dots + (1-f_n)m_n$$

with $m_i \in M_i$. Clearly $f_n(m_n) = 0$ and $m_n = f_{n-1} \dots f_2 f_1(a)$, hence $f_n \dots f_2 f_1(a) = 0$.

Let E, D denote the subsets of $\mathbb{N}$ consisting of all even, odd numbers respectively. Then it is obvious that

$$M = M^*(E) \oplus M(D) = M(E) \oplus M^*(D).$$

Using the second decomposition to complement the summand $M^*(E)$ we get

$$M = M^*(E) \oplus M(E') \oplus M^*(D'),$$

where $E' \subseteq E$ and $D' \subseteq D$.

First assume that all the f_i are monomorphisms. We claim that $D' = D$. Suppose not, and let $t+1$ be the least element in $D - D'$. For any $x \in M_{t+1}$

$$x = \Sigma_1 + (1-f_t)m_t + \Sigma_2$$

where $\Sigma_1 \in M_1 \oplus \dots \oplus M_t$ and $\Sigma_2 \in M_{t+2} \oplus M_{t+3} \oplus \dots$. Consequently $x = -f_t m_t$, thus $f_t : M_t \longrightarrow M_{t+1}$ is onto and hence an isomorphism; which is a contradiction. This proves that $D = D'$, so

$$M = M^*(\mathbb{N}) \oplus M(E').$$

Assume that $E' \neq \emptyset$ and let k be the least element in E'. If E' contains an element $t \neq k$, then by (c) above

$$M = M_1 \oplus \dots \oplus M_k \oplus \dots \oplus M_t \oplus \left(\bigoplus_{i=t}^{\infty} M_i^*\right) \oplus M_k \oplus M(E'')$$

where $E'' = E' - \{k,t\}$. However, this contradicts the fact that the sum is direct. Therefore E' contains exactly one element k. Thus

$$M = M_1 \oplus \dots \oplus M_k \oplus (\overset{\infty}{\underset{i=k}{\oplus}} M_i^*).$$

Again consider an element $x \in M_{k+1}$. Then

$$x = y + \sum_{i=k}^{n} (1 - f_i)m_i$$

where $y \in M_1 \oplus \dots \oplus M_k$, $m_i \in M_i$ and $n \in \mathbb{N}$. Since each f_i is a monomorphism, we get $x = -f_k m_k$, and therefore f_k is onto, hence an isomorphism, a contradiction. This proves that $E' = \emptyset$, and hence $M = M^*(\mathbb{N})$. However this implies that lsTn holds, which is a contradiction since all f_i were assumed to be monomorphisms.

It follows that the f_i are not all monomorphisms. Composing maps together, and reindexing, if necessary, we may assume, without loss of generality, that none of the f_i is a monomorphism. Then applying (a) we get $E' = \emptyset$, and hence

$$M = M^*(E) \oplus M^*(D').$$

But then $M = M^*(\mathbb{N})$ by (b). Hence lsTn follows. □

We end this section with yet another property of modules with decompositions complementing summands.

<u>Proposition 2.27</u>. *A module* M *with a decomposition that complements summands has the internal cancellation property if and only if it is directly finite.*

PROOF. Any module with the internal cancellation property is directly finite (see the proof of Theorem 1.29).

Conversely assume that M is directly finite, and let

$$M = A_1 \oplus B_1 = A_2 \oplus B_2, \text{ with } A_1 \cong A_2.$$

Let $M = \bigoplus_{i \in I} M_i$ be a decomposition that complements summands. There exist subsets J_1, J_2, K_1 and K_2 of I such that

$$J_1 \cap K_1 = \emptyset, J_2 \cap K_2 = \emptyset, I = J_1 \cup K_1 = J_2 \cup K_2;$$

$$A_1 \cong M(J_1), B_1 \cong M(K_1), A_2 \cong M(J_2), B_2 \cong M(K_2).$$

Since M is directly finite, every isotype appears with finite multiplicity. Then $A_1 \cong A_2$ implies that every isotype appears with the same multiplicity in B_1 and B_2. Hence $B_1 \cong B_2$ follows. □

4. INTERNAL CANCELLATION PROPERTY

By Theorem 2.22 and Proposition 2.27, a quasi–continuous directly finite module, which is the direct sum of indecomposable submodules, has the internal cancellation property. In this section, we prove that this result holds for arbitrary quasi–continuous directly finite modules.

Lemma 2.28. *Let* M *be a quasi–continuous module. Then*

(1) M *is purely infinite if and only if* E(M) *is so;*

(2) M *is directly finite if and only if* E(M) *is so.*

PROOF. (1) If M is purely infinite, then $M \cong M \oplus M$, hence $E(M) \cong E(M) \oplus E(M)$, so E(M) is purely infinite.

Conversely, assume that E(M) is purely infinite. Then $E(M) = E_1 \oplus E_2$ with $E(M) \cong E_1 \cong E_2$. As M is quasi–continuous, $M = M_1 \oplus M_2$ where $M_1 = M \cap E_1$ and $M_2 = M \cap E_2$ (Theorem 2.8). Now M_1 and M_2 are relatively injective by Proposition 2.10, and $E(M_1) \cong E(M_2)$. It then follows by Corollary 1.16 that $M_1 \cong M_2$, thus M_1 is quasi–injective. Then M and M_1 are relatively injective by Propositions 1.5 and 1.6. Another application of Corollary 1.16 yields $M_1 \cong M$. Hence M is purely infinite.

(2) If M is not directly finite, then $M \cong M \oplus X$ with $X \neq 0$. Hence $E(M) \cong E(M) \oplus E(X)$ with $E(X) \neq 0$, so E(M) is not directly finite.

Conversely, assume that E(M) is not directly finite. By Theorem 1.35, $E(M) = D \oplus P$ where D is directly finite and P is purely infinite. Again the quasi–continuity of M implies $M = N_1 \oplus N_2$ with $N_1 = M \cap D$ and $N_2 = N \cap P \neq 0$ (since $P \neq 0$). As N_2 is quasi–continuous with purely infinite injective hull, N_2 is purely infinite by (1). Then

$$M = N_1 \oplus N_2 \cong N_1 \oplus N_2 \oplus N_2 = M \oplus N_2,$$

with $N_2 \neq 0$. Therefore M is not directly finite. □

The following is an immediate consequence of the previous lemma and of Theorems 1.35 and 2.8.

Theorem 2.29. *Any quasi–continuous module* M *has a decomposition, unique up to superspectivity,* $M = D \oplus P$, *where* D *is directly finite,* P *is purely infinite, and* P *and* D *are orthogonal.* □

Lemma 2.30. *Any two closures of a submodule* A *of a quasi–continuous module* M *are superspective.*

PROOF. Let M_1 and M_2 be two closures of A. Then $M_1, M_2 \subset^{\oplus} M$ by Proposition 2.4. Write $M = M_1 \oplus X$. Since $A \cap X = 0$, $M_2 \cap X = 0$ and hence $M_2 \oplus X \subset^{\oplus} M$ by (C_3). However $M_2 \oplus X \leq^e M$ since it contains $A \oplus X$. Hence $M = M_2 \oplus X$. □

Theorem 2.31. *Let* M_1 *and* M_2 *be summands of a quasi–continuous module* M. *If* $E(M_1) \cong E(M_2)$, *then* $M_1 \cong M_2$.

PROOF. Let $U = M_1 \cap M_2$ and let X_i be a complement of U in M_i. Then $X_i \subset^{\oplus} M$, $i = 1,2$. One can easily check that the sum $X_1 + U + X_2$ is direct. Then (C_3) implies $M = X_1 \oplus B \oplus X_2 \oplus M^*$ where $U \leq^e B$. Let B_i be a closure of U in M_i, $i = 1,2$. As $U \oplus X_i \leq^e M_i$, $M_i = B_i \oplus X_i$. Write $V_i = B \oplus X_i$. By Lemma 2.30 $B_i \cong B$, hence $V_i \cong M_i$, $i = 1,2$. Therefore our proof will be complete if we show $V_1 \cong V_2$.

By Theorem 2.29, $B = D \oplus P$ where D is directly finite and P is purely infinite. Now X_1, X_2 and P are relatively injective in pairs by Proposition 2.10, and P is quasi–injective by Corollary 2.12. Then $X_1 \oplus P$ and $X_2 \oplus P$ are relatively injective, by Propositions 1.5 and 1.6. Now

$$E(D) \oplus E\,(P \oplus X_1) = E(V_1) \cong E(M_1) \cong E(V_2) = E(D) \oplus E(P \oplus X_2).$$

Since E(D) is directly finite by Lemma 2.28, we get by Theorem 1.29

$$E(P \oplus X_1) \cong E(P \oplus X_2)$$

Hence $P \oplus X_1 \cong P \oplus X_2$ by Corollary 1.16. Therefore

$$V_1 = D \oplus P \oplus X_1 \cong D \oplus P \oplus X_2 = V_2.$$ □

Corollary 2.32. *In a quasi–continuous module* M, *isomorphic submodules have isomorphic closures.*

PROOF. Let $A_i \leq M$ and C_i be a closure of A_i (i= 1,2). If $A_1 \cong A_2$, then $E(C_1) \cong E(C_2)$. Since $C_1, C_2 \subset^{\oplus} M$, $C_1 \cong C_2$ by Theorem 2.31. □

<u>Theorem 2.33</u>. *In a quasi–continuous module* M, *isomorphic directly finite submodules have isomorphic complements. In particular* M *has the internal cancellation property if and only if* M *is directly finite.*

PROOF. Let A_1 and A_2 be directly finite isomorphic submodules of M. Let B_i be a complement of A_i and C_i a complement of B_i which contains A_i, $i = 1,2$. Then $M = C_1 \oplus B_1 = C_2 \oplus B_2$ by Theorem 2.9. Since $E(C_i) = E(A_i)$, $E(C_i)$ is directly finite by Lemma 2.28. Since

$$E(C_1) \oplus E(B_1) = E(M) = E(C_1) \oplus E(B_2),$$

$E(B_1) \cong E(B_2)$ by Theorem 1.29; hence $B_1 \cong B_2$ by Theorem 2.31.

The last statement is obvious. □

A number of applications of Corollary 2.32 will be discussed in Chapter 3.

5. QUASI–CONTINUITY VERSUS QUASI–INJECTIVITY

A quasi–continuous module which is a square is quasi–injective (Corollary 2.11). In generalization of this observation we show that any quasi–continuous module decomposes into a square free and a quasi–injective summand.

<u>Definitions 2.34</u>. A module S is called a *square* if $S \cong X^2$ for some module X; a module is called *square free* if it does not contain a non–zero square.

A submodule T of a module M is called a *square root in* M, if T^2 embeds in M; we say that M is *square full*, if every non–zero submodule of M contains a non–zero square root in M.

<u>Proposition 2.35</u>. *A square full module* M *is quasi–injective if and only if* M *is quasi–continuous.*

PROOF. Any quasi–injective module is continuous (Proposition 2.1).

Conversely assume that M is quasi–continuous. By Zorn's Lemma M contains a direct sum $K = \bigoplus_{\alpha \in \Lambda} S_\alpha$ maximal such that each S_α is a square. Let $S_\alpha \cong X_\alpha^2$, $\alpha \in \Lambda$. Then

$$K \cong \bigoplus_{\alpha \in \Lambda} X_\alpha^2 \cong (\bigoplus_{\alpha \in \Lambda} X_\alpha)^2.$$

Hence $K = K_1 \oplus K_2$ with $K_i \cong \bigoplus_{\alpha \in \Lambda} X_\alpha$, $i = 1,2$. Since M is quasi–continuous, $M = F \oplus N_1 \oplus N_2$ with $K_i \leq^e N_i$, $i = 1,2$ (Lemma 2.21). Then maximality of K implies F is square free. Also $N_1 \cong N_2$ by Corollary 2.32, hence $N_1 \oplus N_2$ is quasi–injective by Corollary 2.11. Since F and $N_1 \oplus N_2$ are relatively injective, in view of Proposition 1.17, M is quasi–injective if and only if F is quasi–injective. We shall prove this by showing that F embeds into N_1 (Corollary 2.11).

To this end we apply Zorn's Lemma to find a monomorphism $F \geq H \xrightarrow{\varphi} N_1$ which cannot be extended. Since N_1 is F–injective, φ can be extended to a homomorphism $\psi : F \longrightarrow N_1$; and it is clear that the restriction of ψ to any closure of H is a monomorphism. Then maximality of the pair (H, φ) implies H is a closed submodule of F and so $F = H \oplus H'$ for some $H' \leq H$. The proof will be complete if we show $H' = 0$.

Assume that $H' \neq 0$. Then H' contains a non–zero square root T. Then $M \geq W = W_1 \oplus W_2$ with $W_1 \cong W_2 \cong T$. Now $W \cap (N_1 \oplus N_2) = 0$ would imply that W embeds in F which is a contradiction, as F is square free. Thus $W \cap (N_1 \oplus N_2) \neq 0$. Applying Lemma 1.31 twice we get that N_1 and W_1 have non–zero isomorphic submodules. Hence we have a non–zero monomorphism $H' \geq H'' \xrightarrow{\theta} N_1$. Let $Y = \varphi H \cap \theta H''$. Since

$$H \geq \varphi^{-1} Y \cong Y \cong \theta^{-1} Y \leq H'' \leq H'$$

and F is square free, $Y = 0$. Then $H \oplus H' \xrightarrow{\varphi \oplus \theta} N_1$ is a monomorphism, and this contradicts the maximality of the pair (H, φ). Hence $H' = 0$ as claimed. □

The conclusion of the following lemma was stated, without proof, for injective modules, in Section 4 of Chapter 1. We include a proof here for the general case.

Lemma 2.36. *Let $\mathcal{A}$ and $\mathcal{B}$ be an orthogonal pair of classes of modules:*

(1) *If a module* M *has* (C_1), *then* $M = A \oplus B$ *with* $A \in \mathcal{A}$ *and* $B \in \mathcal{B}$.

(2) *If* M *is quasi–continuous, then the decomposition in* (1) *is unique up to superspectivity.*

PROOF. (1) By Zorn's Lemma, M has a submodule A maximal with the property $A \in \mathcal{A}$. Since $\mathcal{A}$ is closed under essential extensions, A is a closed submodule of M; hence $A \subset^{\oplus} M$ by (C_1). Write $M = A \oplus B$. Applying the same argument to B, we get $B = C \oplus D$ where C is maximal such that $C \in \mathcal{B}$. Assume that $D \neq 0$. Since $D \notin \mathcal{B}$, D contains a non–zero submodule $Z \in \mathcal{A}$, which is a contradiction to the maximality of A. Hence $D = 0$ and so $M = A \oplus B$ with $A \in \mathcal{A}$ and $B \in \mathcal{B}$.

(2) Let $M = A_1 \oplus B_1 = A_2 \oplus B_2$ with $A_i \in \mathcal{A}$ and $B_i \in \mathcal{B}$, $i = 1,2$. Assume that $M = A_1 \oplus X$. Then $X \cong B_1$, hence $X \in \mathcal{B}$ and therefore $A_2 \cap X = 0$. By (C_3), $A_2 \oplus X \subset^{\oplus} M$, and so $M = A_2 \oplus X \oplus Y$. Then $A_2 \oplus Y \cong A_1$ and $X \oplus Y \cong B_2$. Consequently $Y \in \mathcal{A} \cap \mathcal{B} = 0$, so $M = A_2 \oplus X$. This proves that A_1 and A_2 are superspective. Similarly one can prove that B_1 and B_2 are superspective.

□

<u>Theorem 2.37</u>. *A quasi–continuous module* M *has a decomposition, unique up to superspectivity,* $M = M_1 \oplus M_2$, *where* M_1 *is square free,* M_2 *is square full, and* M_1 *and* M_2 *are orthogonal. Moreover,* M_2 *is quasi–injective.*

PROOF. Consider the hereditary class $\mathcal{X} = \{X : X^2 \rightarrowtail M\}$. Then by Lemma 2.36, M has a decomposition, unique up to superspectivity, $M = M_1 \oplus M_2$ where M_1 is $\mathcal{X}$–void and M_2 is $\mathcal{X}$–full (see Section 4 of Chapter 1). By construction M_1 and M_2 are orthogonal, and it is clear that M_1 is square free. We prove that M_2 is square full. Let N be a non–zero submodule of M_2. Since M_2 is $\mathcal{X}$–full, N contains a non–zero submodule T such that $T^2 \rightarrowtail M$. Since M_1 is $\mathcal{X}$–void, an application of Lemma 1.31 yields $T^2 \rightarrowtail M_2$. Hence M_2 is square full.

Next consider any decomposition $M = A \oplus B$ with the given properties. It is obvious that B is $\mathcal{X}$–full. Now assume that A has a non–zero submodule $X \in \mathcal{X}$. Since A is square free, $X^2 \cap B \neq 0$; otherwise $X^2 \rightarrowtail A$, a contradiction. It then follows by Lemma 2.31 that A and B have non–zero isomorphic submodules, a contradiction. This proves that A is $\mathcal{X}$–void. Therefore the decomposition is unique up to superspectivity. □

COMMENTS

In the context of his investigations of continuous geometries, von Neumann ([36a], [36b], [36c]) introduced regular rings. He called such a ring continuous if its lattice of principal right ideals is upper and lower continuous, indecomposable and infinite dimensional.

Utumi ([60], [61], [66]) studied regular rings in which the lattice of principal right ideals is upper continuous. He called them right continuous regular rings, and characterized them as regular rings with the condition (C_1).

Utumi [65] proceeded to investigate right continuous rings which are not regular; he defined right continuity via the two conditions (C_1) and (C_2). Note that (C_2) is automatically satisfied if the ring is regular. He also utilized (C_3), which is a consequence of (C_2), and a condition (C): for any two right ideals A_1 and A_2 with $A_1 \cap A_2 = 0$, the projection $A_1 \oplus A_2 \longrightarrow\!\!> A_1$ is given by left multiplication by a ring element.

The concepts of continuity and quasi–continuity were generalized to modules by Jeremy [74] and Mohamed and Bouhy [77], the condition (C) was extended to modules by Goel and Jain [78], and was called π–injectivity. It is equivalent to quasi–continuity, and also to the finite extending property defined by Harada [82b].

Theorem 2.8 comprises results of Jeremy [74] and Goel and Jain [78]. The relative injectivity in Proposition 2.10 was observed for continuous rings by Utumi [65], for continuous modules by Mohamed and Bouhy [77], and for quasi–continuous modules by Goel and Jain [78].

The results 2.13, 2.22, 2.24, 2.29, 2.31, 2.33 and special cases of 2.13 are due to Müller and Rizvi ([83], [84]). Theorem 2.13, in full generality, is new, and was demonstrated independently by Mucke [88]. The (easy) special case $M_1 \cap M_2 = 0$ of Theorem 2.31 was observed in Jeremy [74] and in Goel and Jain [78].

Proposition 2.18 and Theorem 2.19 are due to Okado [84]. The material in Section 5 is due to the authors [88b], except that something like Theorem 2.37 is mentioned without proof in Jeremy [74].

The condition (C_1) was studied by Kamal [86], and by Kamal and Müller [88 a,b,c]. Several variations of that condition are investigated in numerous papers by Harada and his collaborators, under the heading of "extending properties". Rings with (C_1) are considered by Chatters et al. ([77], [80]).

CHAPTER 3

CONTINUOUS MODULES

In this chapter we study the structure of the endomorphism rings of continuous and quasi–injective modules. Though many of the basic lemmas hold for quasi–continuous modules, the endomorphism ring of a continuous module M possesses some crucial properties which fail we only assume that M is quasi–continuous.

As an application of these results, in conjunction with some theorems proved in previous chapters, we show that continuous modules have the exchange property.

Beyond these facts involving the endomorphism ring, we will discuss a few other properties of continuous modules, which do not generally hold for quasi–continuous modules.

1. ENDOMORPHISM RINGS

Throughout this section, S will denote the endomorphism ring of a module M, J the Jacobson radical of S, $\Delta = \{\alpha \in S : \operatorname{Ker} \alpha \leq^{e} M\}$ and $\overline{S} = S/\Delta$.

The following lemma, whose proof is straightforward, will be used freely in this section.

<u>Lemma 3.1</u>. *Let* A *be a submodule of* M, $\alpha \in S$, *and* e *an idempotent of* S. *Then:*

(1) *If* $A \leq^{e} M$, *then* $eA \leq^{e} eM$;

(2) $\alpha M \leq eM$ *if and only if* $\alpha S \leq eS$. □

<u>Lemma 3.2</u>. *For an arbitrary module* M,

(1) Δ *is an ideal; and*

(2) *if* $\{e_i : i \in I\}$ *is a family of idempotents of* S *which are orthogonal* modulo Δ, *then* $\sum_{i \in I} e_i M$ *is direct.*

PROOF. (1) Let a, b $\in \Delta$ and $\alpha \in S$. Then $\operatorname{Ker} a \leq^{e} M$ and $\operatorname{Ker} b \leq^{e} M$. Since $\operatorname{Ker}(a-b) \geq \operatorname{Ker} a \cap \operatorname{Ker} b$ and $\operatorname{Ker} \alpha a \geq \operatorname{Ker} a$, $\operatorname{Ker}(a-b)$ and $\operatorname{Ker} \alpha a$ are essential submodules of M; and consequently, $a-b \in \Delta$ and $\alpha a \in \Delta$. Let $N = \{n \in M : \alpha(n) \in \operatorname{Ker} a\}$. Then it is clear that $N \leq^{e} M$ and $\operatorname{Ker} a\alpha \geq N$. Hence $a\alpha \in \Delta$.

(2) It suffices to consider a finite family e_i. For $i \neq j$, $e_ie_j \in \Delta$ and hence $\text{Ker } e_ie_j \leq^e M$. Since a finite intersection of essential submodules is again essential, there exists an essential submodule K of M such that $e_ie_jK = 0$ holds for all $i \neq j$. It follows immediately that $\Sigma\, e_iK$ is direct. But $e_iK \leq^e e_iM$, and consequently $\Sigma\, e_iM$ is also direct. □

<u>Lemma 3.3</u>. *Let* $M = M_1 \oplus M_2$. *If* M_1 *and* M_2 *are orthogonal, then* $S/\Delta \cong S_1/\Delta_1 \times S_2/\Delta_2$. *The converse holds if* M_1 *and* M_2 *are relatively injective.*

PROOF. We can write any $s \in S$ as $s = \begin{bmatrix} s_1 & \psi \\ \varphi & s_2 \end{bmatrix}$ where $s_1 \in \text{End } M_1$, $s_2 \in \text{End } M_2$, $\varphi \in \text{Hom}(M_1, M_2)$ and $\psi \in \text{Hom}(M_2, M_1)$; further s_1, s_2, φ and ψ may be considered as elements of S by defining them to be zero on the other summand. Then orthogonality of M_1 and M_2 implies $\varphi, \psi \in \Delta$. It is also clear that $\text{Ker } s \cap M_1 = \text{Ker } s_1 \cap \text{Ker } \varphi$ and $\text{Ker } s \cap M_2 = \text{Ker } s_2 \cap \text{Ker } \psi$.

Now we prove that $s \in \Delta$ if and only if $s_1 \in \Delta_1$ and $s_2 \in \Delta_2$. Assume that $s \in \Delta$. Then $\text{Ker } s \leq^e M$ and hence $\text{Ker } s_1 \cap \text{Ker } \varphi = \text{Ker } s \cap M_1 \leq^e M_1$. But then $\text{Ker } s_1 \leq^e M_1$ and $s_1 \in \Delta_1$. Similarly $s_2 \in \Delta_2$. Conversely, assume that $s_1 \in \Delta_1$ and $s_2 \in \Delta_2$. Since $\text{Ker } \varphi \leq^e M_1$, $\text{Ker } \varphi \cap \text{Ker } s_1 \leq^e M_1$, and hence $\text{Ker } s \cap M_1 \leq^e M_1$. Similarly $\text{Ker } s \cap M_2 \leq^e M_2$. Therefore $\text{Ker } s \leq^e M$ and $s \in \Delta$. Hence

$$S/\Delta = \begin{bmatrix} S_1/\Delta_1 & 0 \\ 0 & S_2/\Delta_2 \end{bmatrix} \cong S_1/\Delta_1 \times S_2/\Delta_2.$$

The last statement is obvious. □

A ring is called *reduced* if it has no non–zero nilpotent elements; every idempotent of a reduced ring is central (cf. Stenstrom [75], p. 40).

<u>Lemma 3.4</u>. *If* M *is square free, then* $\overline{S}$ *is reduced. In particular, all idempotents of* $\overline{S}$ *are central.*

PROOF. Let $\alpha \in S$ such that $\alpha^2 \in \Delta$. Let $K = \text{Ker } \alpha^2$ and L be a complement of $\text{Ker } \alpha$. Then $K \leq^e M$ and $\text{Ker } \alpha \oplus L \leq^e M$. Now

$$\text{Ker } \alpha \geq \alpha(K \cap L) \cong K \cap L.$$

Since M is square free, $K \cap L = 0$ and consequently $L = 0$. Therefore $\text{Ker } \alpha \leq^e M$ and $\alpha \in \Delta$. □

<u>Proposition 3.5</u>. *If* M *is continuous, then* S/Δ *is a* (von Neumann) *regular ring and* Δ *equals* J, *the Jacobson radical of* S.

PROOF. Let $\alpha \in S$ and let L be a complement of $K = \operatorname{Ker} \alpha$. By (C_1), $L \subset^{\oplus} M$. Since $\alpha|_L$ is a monomorphism, $\alpha L \subset^{\oplus} M$ by (C_2). Hence there exists $\beta \in S$ such that $\beta\alpha = 1_L$. Then

$$(\alpha - \alpha\beta\alpha)\,(K \oplus L) = (\alpha - \alpha\beta\alpha)L = 0,$$

and so $K \oplus L \leq \operatorname{Ker}(\alpha - \alpha\beta\alpha)$. Since $K \oplus L \leq^e M$, $\alpha - \alpha\beta\alpha \in \Delta$. Therefore S/Δ is a regular ring. This also proves that $J \leq \Delta$.

Let $a \in \Delta$. Since $\operatorname{Ker} a \cap \operatorname{Ker}(1-a) = 0$ and $\operatorname{Ker} a \leq^e M$, $\operatorname{Ker}(1-a) = 0$. Hence $(1-a)M \subset^{\oplus} M$ by (C_2). However $(1-a)M \leq^e M$ since $\operatorname{Ker} a \leq (1-a)M$. Thus $(1-a)M = M$, and therefore $1-a$ is a unit in S. It then follows that $a \in J$, and hence $\Delta \leq J$. □

<u>Remark</u>. The conclusion of the previous proposition fails for quasi–continuous modules (see Proposition 3.15).

<u>Lemma 3.6</u>. *If* M *is quasi–continuous, then* $\overline{S}_{\overline{S}}$ *has* (C_3).

PROOF. By Theorem 2.37, $M = M_1 \oplus M_2$ where M_1 is square free, M_2 is quasi–injective, and M_1 and M_2 are orthogonal. Then $\overline{S} = \overline{S}_1 \times \overline{S}_2$ by Lemma 3.3. Since M_2 is continuous, $\overline{S}_2$ is regular by Proposition 3.5, hence $\overline{S}_2$ has (C_2). Let $\overline{e}$, $\overline{f}$ be idempotents of $\overline{S}_1$ such that $\overline{eS}_1 \cap \overline{fS}_1 = 0$. Since $\overline{e}$ and $\overline{f}$ are central by Lemma 3.4, $\overline{ef} = \overline{fe} \in \overline{eS}_1 \cap \overline{fS}_1 = 0$. Thus $\overline{e}$ and $\overline{f}$ are orthogonal idempotents, and $\overline{eS}_1 \oplus \overline{fS}_1$ is a summand of $\overline{S}_1$. Hence $\overline{S}_{1\overline{S}_1}$ satisfies (C_3). Therefore $\overline{S}_{\overline{S}}$ satisfies (C_3). □

<u>Lemma 3.7</u>. *If* M *is quasi–continuous, then idempotents* modulo Δ *can be lifted.*

PROOF. Consider $\varepsilon \in S$ such that $\varepsilon^2 - \varepsilon \in \Delta$, and let $K = \operatorname{Ker}(\varepsilon^2 - \varepsilon)$. Since $\varepsilon K \cap (1 - \varepsilon)K = 0$, $M = M_1 \oplus M_2$ such that $\varepsilon K \leq M_1$ and $(1 - \varepsilon)K \leq M_2$. Let e be the projection $M_1 \oplus M_2 \longrightarrow\!\!> M_1$. Then

$$(e - \varepsilon)K \leq (e - \varepsilon)\,\varepsilon K + (e - \varepsilon)(1 - \varepsilon)K = 0.$$

Since $K \leq^e M$, $e - \varepsilon \in \Delta$. □

Lemma 3.8. *Let* M *be a quasi–continuous module, and* $\{e_i : i \in I\}$ *a family of idempotents in* S. *Then the following are equivalent:*

(1) $\sum_{i\in I} e_i M$ *is direct;*

(2) *There exist orthogonal idempotents* f_i *such that* $e_iS = f_iS$ *for every* $i \in I$;

(3) $\Sigma\, \bar{e}_i\bar{S}$ *is direct.*

PROOF. (2) ⇒ (3): Trivial.

(3) ⇒ (1): It is enough to consider a finite family e_i. Since $\bar{S}_{\bar{S}}$ has (C_3) and $\Sigma\, \bar{e}_i\bar{S}$ is direct, we get $\oplus\, \bar{e}_i\bar{S} \subset^{\oplus} \bar{S}$. Thus there exist orthogonal idempotents $\bar{g}_i$ of $\bar{S}$ such that $\bar{e}_i\bar{S} = \bar{g}_i\bar{S}$. In view of Lemma 3.7, we may assume that the g_i are idempotents of S. Then $\Sigma\, g_iM$ is direct by Lemma 3.2. Now $\bar{e}_i = \bar{g}_i\bar{e}_i$ and so $e_i - g_ie_i \in \Delta$. Hence there exist essential submodules K_i such that $(e_i - g_ie_i)K_i = 0$. It is clear that $e_iK_i \leq g_iM$,, and hence $\Sigma\, e_iK_i$ is direct. But since $e_iK_i \leq^e e_iM$, $\Sigma\, e_iM$ is direct.

(1) ⇒ (2): Let C_i be a closure of $\sum_{j\neq i} e_jM$. Since M is quasi–continuous,

$$M = e_iM \oplus C_i \oplus D_i$$

for some $D_i \leq M$. It is easy to check that $e_iM + C_i + (1 - e_i)\, D_i$ is direct; and since $D_i \leq e_i\, D_i \oplus (1 - e_i)D_i$,

$$M = e_iM \oplus C_i \oplus (1-e_i)D_i.$$

Let f_i be the projection of M onto e_iM with kernel $C_i \oplus (1-e_i)D_i$. Then $f_i^2 = f_i$ and $e_iM = f_iM$. Hence $e_iS = f_iS$, $e_if_i = f_i$ and $f_ie_i = e_i$. Since $e_jM \leq C_i$ for every $j \neq i$, $f_ie_j = 0$. Thus $f_if_j = f_ie_jf_j = 0.$ for all $j \neq i$. Therefore, the f_i are orthogonal, and (2) holds. □

Now assume that the idempotents e_i are orthogonal modulo Δ. Then of course $\Sigma\, \bar{e}_i\bar{S}$ is direct, and hence Lemma 3.8 applies. In this situation, we show that the f_i constructed in (2) satisfy $\bar{f}_i = \bar{e}_i$. First we note that $e_i - f_i$ vanishes on $e_i\, M \oplus (1-e_i)D_i$. Also $(e_i - f_i)e_j = e_ie_j$, for $j \neq i$. Hence $(e_i - f_i)e_j \in \Delta$, and therefore there exist essential submodules N_j such that $e_jN_j \leq \operatorname{Ker}(e_i - f_i)$, for $j \neq i$. Since $e_jN_j \leq^e e_jM$,

$$\bigoplus_{j \neq i} e_j N_j \leq^e \bigoplus_{j \neq i} e_j M \leq^e C_i.$$

This all shows that $\operatorname{Ker}(e_i - f_i) \leq^e M$, and hence $e_i - f_i \in \Delta$.

This, along with Lemma 3.7, proves the following:

Corollary 3.9. *If* M *is quasi–continuous, then any family of orthogonal idempotents of* $\overline{S}$ *lifts to a family of orthogonal idempotents of* S. □

Theorem 3.10. *If* M *is quasi–injective, then* $\overline{S}$ *is right self–injective and regular.*

PROOF. $\overline{S}$ is regular by Proposition 3.5. Let $\overline{A}$ be a right ideal of $\overline{S}$, and $\varphi : \overline{A} \longrightarrow \overline{S}$ a homomorphism. We use Zorn's Lemma to obtain a maximal direct sum $\bigoplus_{i \in I} \overline{e}_i\overline{S}$ of principal right ideals contained in $\overline{A}$. Then $\bigoplus_{i \in I} \overline{e}_i\overline{S} \leq^e \overline{A}$. Since $\overline{S}$ is regular and idempotents modulo Δ lift by Lemma 3.7, we may assume that each e_i is an idempotent of S. Then $\sum_{i \in I} e_i M$ is direct by Lemma 3.8.

Let $\varphi(\overline{e}_i) = \overline{x}_i = \overline{x}_i\overline{e}_i$. Define $\psi_i : e_iM \longrightarrow M$ by $\psi_i(e_im) = x_ie_im$, and $\psi : \bigoplus_{i \in I} e_iM \longrightarrow M$ by $\psi = \bigoplus_{i \in I} \psi_i$. Since M is quasi–injective, ψ extends to an element $\alpha \in S$. Then

$$0 = (\alpha - \psi_i)e_i = (\alpha - x_ie_i)e_i,$$

and hence

$$0 = (\overline{\alpha} - \overline{x}_i\overline{e}_i)\overline{e}_i = (\overline{\alpha} - \varphi)\overline{e}_i.$$

Now let $\overline{a} \in \overline{A}$. There exists an essential right ideal $\overline{K}$ of $\overline{S}$ such that $\overline{aK} \leq \bigoplus_{i \in I} \overline{e}_i\overline{S}$. Then

$$(\overline{\alpha} - \varphi)(\overline{a})\overline{K} = (\overline{\alpha} - \varphi)(\overline{aK}) \leq (\overline{\alpha} - \varphi)\,(\bigoplus_{i \in I} \overline{e}_i\overline{S}) = 0.$$

Since $\overline{S}$, being regular, is non–singular, we get $(\overline{\alpha} - \varphi)(\overline{a}) = 0$. Thus $\overline{\alpha}(\overline{a}) = \overline{\alpha}(\overline{a}) = \varphi(\overline{a})$ for all $\overline{a} \in \overline{A}$, and $\overline{\alpha}$ extends φ. □

Theorem 3.11. *If* M *is continuous, then* $\overline{S}$ *is right continuous and regular.*

PROOF. Let $\overline{A}$ be a right ideal of $\overline{S}$. As in the proof of the previous theorem, there exist idempotents $e_i \in S$ such that $\bigoplus_{i \in I} \overline{e}_i\overline{S} \leq^e \overline{A}$. Let eM be a closure of $\bigoplus_{i \in I} e_iM$. We claim that $\bigoplus_{i\in I} \overline{e}_i\overline{S} \le^e \overline{eS}$. Assume that $(\bigoplus_{i\in I} \overline{e}_i\overline{S}) \cap \overline{\mathscr{E}S} = 0$ for some $\overline{\mathscr{E}} \in \overline{eS}$. Since $\overline{S}$ is regular and idempotents lift modulo Δ, we may assume that $\mathscr{E}$ is an idempotent in S. Then $(\bigoplus_{i\in I} e_iM) \cap \mathscr{E}M = 0$ by Lemma 3.8. Now $\overline{e\mathscr{E}} = \overline{\mathscr{E}}$, hence $e\mathscr{E} - \mathscr{E} \in \Delta$ and consequently $(e\mathscr{E} - \mathscr{E})K = 0$ for some $K \le^e M$. If follows that $\mathscr{E}K \le eM$, so $\mathscr{E}K = 0$ since $\bigoplus_{i\in I} e_i M \le^e eM$. Thus $\mathscr{E} \in \Delta$ and $\overline{\mathscr{E}} = 0$, proving our claim.

Let $\overline{a}$ be an arbitrary element in $\overline{A}$. Then

$$\overline{a}\,\overline{S} \cap (\bigoplus_{i\in I} \overline{e}_i\,\overline{S}) \le^e \overline{a}\,\overline{S} \cap \overline{A} = \overline{aS},$$

and

$$\overline{aS} \cap (\bigoplus_{i\in I} \overline{e}_i\overline{S}) \le \overline{aS} \cap \overline{eS} \le \overline{aS}.$$

Hence $\overline{aS} \cap \overline{eS} \le^e \overline{aS}$. However $\overline{aS} \cap \overline{eS}$ is generated by an idempotent since $\overline{S}$ is regular. Thus $\overline{aS} \cap \overline{eS} = \overline{aS}$ and consequently $\overline{aS} \le \overline{eS}$. Therefore $\overline{A} \le \overline{eS}$. Since $\bigoplus_{i\in I} \overline{e}_i\overline{S} \le \overline{A}$, $\overline{A} \le^e \overline{eS}$.

□

We recall that any quasi–continuous module has a decomposition into orthogonal summands where one is square free and the other square full (Theorem 2.37). The following proposition provides an alternate characterization of these summands.

<u>Proposition 3.12</u>. *Let* M *be a quasi–continuous module. Then*

(1) M *is square free if and only if* $\overline{S}$ *is reduced;*

(2) M *is square full if and only if every non–zero right ideal of* $\overline{S}$ *has a non–zero nilpotent element.*

PROOF. If M is square free, then $\overline{S}$ has no non–zero nilpotent elements (Lemma 3.4). Now we prove that if M is square full, then every non–zero right ideal of $\overline{S}$ contains a non–zero nilpotent elememt. Let $\alpha \in S - \Delta$. Since M is quasi–continuous, $M = (\mathrm{Ker}\,\alpha)^* \oplus B$ where $(\mathrm{Ker}\,\alpha)^*$ is a closure of Ker α. Then $B \ne 0$; otherwise Ker α would be essential in M. Since M is square full, there exists $0 \ne Z \le \alpha B$ such that $Z^2 \rightarrowtail M$. Thus $M \ge Z_1 \oplus Z_2$ with $Z \cong Z_1 \cong Z_2$. Define X and Y as follows: If $Z \cap Z_1 = 0$, $X = Z$ and $Y = Z_1$; and if $Z \cap Z_1 \ne 0$, $X = Z \cap Z_1$ and Y the image of X under the isomorphism $Z \longrightarrow Z_2$. In either case we get $0 \ne X \le \alpha B$ and $Y \le M$ such that $X \cap Y = 0$ and $X \cong Y$. Let $\sigma : Y \longrightarrow X$ be an isomorphism, and let X' be a complement of Y which contains X. Define $\mu : X' \oplus Y \longrightarrow M$ by $\mu = 0$ on X' and

$\mu = (\alpha|_B)^{-1}\sigma$ on Y. Since M is quasi-injective by Proposition 2.35, μ exends to $\beta \in S$. Now, for any $y \in Y$, $\sigma(y) \in X \leq \alpha B$, and hence

$$\alpha\beta(y) = \alpha(\alpha|_B)^{-1} \sigma(y) = \sigma(y).$$

Since σ is an isomorphism, Ker $\alpha\beta \cap Y = 0$ and so $\alpha\beta \notin \Delta$. However

$$(\alpha\beta)^2 (X' \oplus Y) = (\alpha\beta)^2 Y = \alpha\beta\,(\alpha\beta Y) = \alpha\beta(\sigma Y) = \alpha\beta X = 0.$$

Hence $(\alpha\beta)^2 \in \Delta$ since $X' \oplus Y \leq^e M$.

So far we have proved the "only if" parts in (1) and (2). Now by Theorem 2.37, $M = M_1 \oplus M_2$ where M_1 and M_2 are orthogonal, M_1 square free and M_2 square full. Then by Lemma 3.3, $\overline{S} = \overline{S}_1 \oplus \overline{S}_2$ where $S_i = \text{End } M_i$, $i = 1,2$. By what we have already proved, $\overline{S}_1$ has no non-zero nilpotent elements and every non-zero right ideal of $\overline{S}_2$ contians a non-zero nilpotent element. If $\overline{S}$ has no non-zero nilpotent elements, then $\overline{S}_2 = 0$, thus $S_2 = \Delta_2$ and hence $S_2 = 0$, consequently $M_2 = 0$. This proves the "if" part of (1). The "if" part of (2) follows similarly. □

<u>Corollary 3.13</u>. *Let M be quasi-continuous. Then there is a ring decomposition* $\overline{S} = \overline{S}_1 \times \overline{S}_2$ *such that* $\overline{S}_1$ *is regular and right self-injective, and* $\overline{S}_2$ *is reduced.*

PROOF. By Theorem 2.37, $M = M_1 \oplus M_2$, where M_1 is square full and quasi-injective, M_2 is square free, and both are orthogonal. Orthogonality implies $\overline{S} = \overline{S}_1 \times \overline{S}_2$ (Lemma 3.3).

By Theorem 3.10, $\overline{S}_1$ is regular and self injective. By Proposition 3.12, $\overline{S}_2$ is reduced. □

<u>Remarks</u>. (1) In the situation of Corollary 3.13, assume that M is non-singular. Then $\Delta = 0$, hence $S = S_1 \times S_2$, and the decomposition $M = M_1 \oplus M_2$ is unique.

(2) If, in (1), $M = R$, then $R = R_1 \times R_2$ where R_1 is regular and right-self-injective, and R_2 is reduced and right-quasi-continuous.

(3) If, in (1), M is continuous, then $\overline{S}_2$ is regular and right-continuous (Theorem 3.11).

2. CONTINUOUS MODULES

In this section we use Corollary 2.32 and results obtained in the previous section, to give necessary and sufficient conditions for a quasi–continuous module to be continuous. We also discuss some properties of continuous modules, which fail to hold for quasi–continuous modules.

Lemma 3.14. *A quasi–continuous module M is continuous if and only if every monomorphism M >——> M with essential image is an isomorphism.*

PROOF. The "only if" part is obvious. Conversely, assume the condition and let $N \subset^{\oplus} M$ and $f : N \longrightarrow M$ be a monomorphism. As M is quasi–continuous, $M = A \oplus B$ with $fN \leq^e B$. Since $fN \cong N$, $B \cong N$ by Corollary 2.32. Let $g : B \longrightarrow N$ be an isomorphism. Then

$$M = A \oplus B \xrightarrow{1\oplus g} A \times N \xrightarrow{1\oplus f} A \oplus B = M$$

is a monomorphism with image $A \oplus fN \leq^e M$. By assumption, $(1 \oplus f)(1 \oplus g)$ is an isomorphism. Hence $1 \oplus f$ is onto, and consequently $B = fN$. Thus $fN \subset^{\oplus} M$ and (C_2) holds. □

Proposition 3.15. *Let M be a quasi–continuous module,* $S = \text{End}\, M$, $\Delta = \{\alpha \in S : \text{Ker}\, \alpha \leq^e M\}$ *and J the Jacobson radical of S. Then M is continuous if and only if* $\Delta = J$ *and* S/Δ *is regular.*

PROOF. Necessity follows by Proposition 3.5. Conversely, assume that $\Delta = J$ and S/Δ is regular. Let $\varphi \in S$ be a monomorphism with essential image. There exists $\psi \in S$ such that $\varphi - \varphi\psi\varphi \in \Delta$. Consequently $(1-\varphi\psi)\varphi K = 0$ for some $K \leq^e M$. Since φ is a monomorphism, $\varphi K \leq^e \varphi M$; thus $\varphi K \leq^e M$ as $\varphi M \leq^e M$. Therefore $1 - \varphi\psi \in \Delta = J$, and hence $\varphi\psi$ is a unit in S. Thus φ is onto, and consequently φ is an isomorphism. Then M is continuous by Lemma 3.14. □

Theorem 3.16. *The following are equivalent, for a module* $M = \bigoplus_{\alpha\in\Lambda} M_\alpha$:

(1) *M is continuous;*

(2) *M is quasi–continuous and the* M_α *are continuous;*

(3) M_α *is continuous and* M_β*–injective for all* $\alpha \neq \beta$, *and* (A_2) *holds.*

PROOF. (1) implies (2) trivially, and (2) and (3) are equivalent by Theorem 2.13. It remains to see that (2) implies (1). According to Lemma 3.14, we have to establish that every essential monomorphism f : M >——> M is onto.

We first deal with the case of a finite index set $\Lambda = \{1, ..., n\}$. Here we have $E(M) = \bigoplus_{\alpha=1}^{n} E(M_\alpha) = \bigoplus_{\alpha=1}^{n} E(fM_\alpha)$. We choose closures C_α of fM_α in M. Since $fM_\alpha \cong M_\alpha \subset^\oplus M$, Corollary 2.32 yields $C_\alpha \cong M_\alpha$. Thus we obtain an essential monomorphism $M_\alpha \cong fM_\alpha \leq^e C_\alpha \cong M_\alpha$ of M_α, which is an isomorphism by Lemma 3.14. Consequently $fM_\alpha = C_\alpha$ is a summand of M, and hence closed. We deduce $M \cap E(fM_\alpha) = fM_\alpha$. As M is quasi–continuous, we infer $M = \bigoplus_{\alpha=1}^{n} M \cap E(fM_\alpha) = \bigoplus_{\alpha=1}^{n} fM_\alpha = fM$. This completes the proof in the finite case.

In the general case, suppose there is an essential monomorphism $f : M >\!\!\longrightarrow M$ which is not onto. Inductively we shall construct a sequence $x_n \in M_{\alpha_n} - fM$, with distinct α_n and strictly increasing annihilators x_n^o. This contradicts (A_3), which is valid for M according to Proposition 2.24.

Let $x_i \in M_{\alpha_i}$ be constructed as claimed, for $i \leq n$. Write $A = \{\alpha_1, ..., \alpha_n\}$. $M(A) \cap fM$ is obviously essential in $M(A)$; therefore $M(A)$ is a closure of $M(A) \cap fM$ in M. As $fM \cong M$ is quasi–continuous, $M(A) \cap fM$ possesses another closure, V, in fM. Now pick a closure W of V in M; then $M(A) \cap fM \leq^e V \leq^e W \subset^\oplus M$. Clearly W is also a closure of $M(A) \cap fM$ in M. By Corollary 2.32, all these closures $M(A)$, V and W are isomorphic. They are continuous, by the finite case established earlier. Thus the inclusion $V \leq^e W$ yields an essential monomorphism $W >\!\!\longrightarrow W$. We conclude $V = W$, by Lemma 3.14.

As $V \cap M(\Lambda - A) = 0$, and both submodules are summands, $V \oplus M(\Lambda - A)$ is a summand of M by (C_3). But $V \geq M(A) \cap fM \leq^e M(A)$ implies that $V \oplus M(\Delta - A)$ is essential in M, and consequently equal to M. Write $x_n = v + \Sigma\, y_i$, accordingly. As $x_n \notin fM$ and $v \in fM$, there exists $y_i \notin fM$. Take x_{n+1} to be such y_i, and α_{n+1} its index i. Clearly $\alpha_{n+1} \neq \alpha_1, ..., \alpha_n$. Moreover, $x_{n+1} = \pi(x_n)$, for the projection $\pi : M = V \oplus M(\Lambda - A) \longrightarrow\!\!\!\!> M_{\alpha_{n+1}}$. Thus $x_n^o \leq x_{n+1}^o$. By essentiality there exists $r \in R$ with $0 \neq x_n r \in M(A) \cap fM \leq V$. We deduce $x_{n+1} r = \pi(x_n r) = 0$, and therefore $r \in x_{n+1}^o - x_n^o$. This completes the construction, and the proof.

□

As yet another application of Corollary 2.32, we show that continuous modules satisfy the conclusion of the Schröder– Bernstein Theorem.

Theorem 3.17. *Let* M *be a continuous module and* N *a quasi–continuous module. If* $M \rightarrowtail N$ *and* $N \rightarrowtail M$, *then* $M \cong N$.

PROOF. Without loss of generality, we may assume $N \leq M$. Let $\varphi : M \rightarrowtail N$ be a monomorphism. Since M is continuous, $\varphi M \subset^{\oplus} M$, and hence $\varphi M \subset^{\oplus} N$. Write $N = A \oplus \varphi M$, and let

$$B = A + \varphi A + \varphi^2 A + \dots$$

(in fact the sum is direct). Then $B = A \oplus \varphi B$. Since φM is (quasi–) continuous, $\varphi M = P \oplus Q$ with $\varphi B \leq^e P$. Then

$$B = A \oplus \varphi B \leq^e A \oplus P.$$

Since $B \cong \varphi B$, $P \cong A \oplus P$ by Corollary 2.32. Thus

$$N = A \oplus \varphi M = A \oplus P \oplus Q \cong P \oplus Q = \varphi M \cong M. \qquad \square$$

Corollary 3.18. *Mutually subisomorphic continuous modules are isomorphic.*

□

Remark. Two subisomorphic quasi–continuous modules need not be isomorphic. For an example, let R be a commutative domain which is not a principal ideal domain (e.g. take R = k[x,y] for some field k). Pick an ideal A of R which is not principal. Since A and R are uniform R–modules, they are trivially quasi–continuous. However $A \leq R \rightarrowtail A$ and $A \ncong R$.

3. THE EXCHANGE PROPERTY

In this section, we establish the exchange property for continuous modules. The proof is based on the decomposition of a (quasi–)continuous module into a direct sum of a quasi–injective and a square free part (Theorem 2.37). For the latter, the fact that idempotents of the endomorphism ring modulo the radical are central, is used to verify a criterion which was provided by Zimmermann–Huisgen and Zimmermann [84].

We start by proving some results concerning the exchange property.

Lemma 3.19. *If* M *has the exchange property and*

$$A = M \oplus N \oplus L = \bigoplus_{i \in I} A_i \oplus L,$$

then there exist submodules $B_i \leq A_i$ *such that* $A = M \oplus (\bigoplus_{i \in I} B_i) \oplus L$.

PROOF. Let p be the projection of M onto $\bigoplus_{i\in I} A_i$ with kernel L. Then the restriction of p to $M \oplus N$ is an isomorphism. Now

$$pM \oplus pN = \bigoplus_{i\in I} A_i.$$

Since pM, being isomorphic to M, has the exchange property, we get

$$pM \oplus pN = pM \oplus (\bigoplus_{i\in I} B_i)$$

with $B_i \leq A_i$. Hence

$$A = M \oplus N \oplus L = p^{-1}(pM \oplus (\bigoplus_{i\in I} B_i)) = M \oplus (\bigoplus_{i\in I} B_i) \oplus L. \qquad \square$$

Lemma 3.20. *Let* $M = X \oplus Y$. *Then* M *has the exchange property if (and only if)* X *and* Y *have the exchange property.*

PROOF. Assume that X and Y have the exchange property; and let

$$A = M \oplus N = \bigoplus_{i\in I} A_i.$$

Then $A = X \oplus Y \oplus N = Y \oplus (\bigoplus_{i\in I} B_i)$ with $B_i \leq A_i$. It then follows by the previous lemma that

$$A = X \oplus Y \oplus (\bigoplus_{i\in I} C_i),$$

with $C_i \leq B_i$. Thus M has the exchange property. The converse is left to the reader.

$\square$

Definition 3.21. Given two modules U and V, a family $(f_i)_{i\in I}$ of homomorphisms $U \longrightarrow V$ is called *summable* if for each $u \in U$, $f_i(u) = 0$ for almost all $i \in I$. (Then Σf_i is a well defined homomorphism $U \longrightarrow V$.)

For a module M, let S, Δ and J be as defined in the first section of this chapter.

Proposition 3.22. *The following are equivalent for a module* M :

(1) M *has the exchange property;*

(2) *If* $M \oplus N = \bigoplus_{i\in I} A_i$, *with* $A_i \cong M$ *for all* $i \in I$, *then there exist submodules* $C_i \leq A_i$ *such that* $M \oplus N = M \oplus (\bigoplus_{i\in I} C_i)$.

(3) *For each summable family* $(f_i)_{i\in I}$ *in* S *with* $\Sigma f_i = 1$, *there exist orthogonal idempotents* $e_i \in Sf_i$ *such that* $\Sigma e_i = 1$.

PROOF. (1) ⇒ (2) is trivial.

(2) ⇒ (3): Let $(f_i)_{i\in I}$ be a summable family of elements of S such that $\Sigma f_i = 1$. Define $A = \bigoplus_{i\in I} A_i$ with $A_i = M$ for all $i \in I$. Define $f : M \longrightarrow A$ by $f(m) = (f_i(m))_{i\in I}$; and $g : A \longrightarrow M$ by $g((m_i)_{i\in I}) = \sum_{i\in I} m_i$. It is clear that $gf = 1_M$, and so $A = fM \oplus \text{Ker } g$. By hypothesis, $A_i = B_i \oplus C_i$ such that

$$A = fM \oplus (\bigoplus_{i\in I} C_i).$$

Let p be the projection of A onto $\bigoplus_{i\in I} B_i$ with kernel $\bigoplus_{i\in I} C_i$. Then the restriction of p to fM is an isomorphism; and it is obvious that $pfg\,p^{-1}$ is the identity on $\bigoplus_{i\in I} B_i$. Let $\pi_j : \bigoplus_{i\in I} B_i \longrightarrow\!\!> B_j$ be the natural projection, and define $e_i = g\,p^{-1}\,\pi_i\,pf$. Then

$$e_i e_j = g\,p^{-1}\,\pi_i\,pf\,g\,p^{-1}\pi_j pf = gp^{-1}\,\pi_i\,\pi_j\,pf.$$

Thus $e_i e_j = 0$ for $j \neq i$ and $e_i^2 = e_i$.

Next, let ρ_i be the projection $B_i \oplus C_i \longrightarrow\!\!> B_i$. For any $m \in M$

$$\pi_i pf(m) = \pi_i p(f_j(m))_{j\in I} = \pi_i(\rho_j f_j(m))_{j\in I} = \rho_i f_i(m).$$

Hence $\pi_i pf = \rho_i f_i$ and consequently $e_i = gp^{-1}\,\rho_i f_i \in Sf_i$. In particular, the family $(e_i)_{i\in I}$ is again summable, and $\Sigma e_i = 1$ follows by construction.

(3) ⇒ (1): Let $X = M \oplus Y = \bigoplus_{i\in I} X_i$. Let $\mu_j : \oplus X_i \longrightarrow\!\!> X_j$ and $q : M \oplus Y \longrightarrow\!\!> M$ denote the natural projections, and define $h_i = q\,\mu_i|_M$. Then clearly $h_i \in S$, the family $(h_i)_{i\in I}$ is summable, and $\Sigma h_i = 1$. By hypothesis, we can find orthogonal idempotents $\varepsilon_i = s_i h_i \in Sh_i$ with $\Sigma\,\varepsilon_i = 1$. Define $\varphi_i : X \longrightarrow M$ by $\varphi_i = \varepsilon_i s_i q \mu_i$. We claim that

$$X = M \oplus (\bigoplus_{i\in I} (X_i \cap \text{Ker } \varphi_i)).$$

Once this is established, (1) follows.

First note that $(\varphi_i)_{i\in I}$ is summable; let $\varphi = \Sigma\,\varphi_i$. Next $\varphi_i|_M = \varepsilon_i$; indeed

$$\varphi_i(m) = \varepsilon_i s_i\,q\,\mu_i\,(m) = \varepsilon_i s_i\,h_i(m) = \varepsilon_i\,\varepsilon_i(m) = \varepsilon_i(m)$$

for every $m \in M$. Hence

$$\varphi|_M = (\Sigma\,\varphi_i)|_M = \Sigma\,\varepsilon_i = 1_M.$$

Thus $X = M \oplus \text{Ker } \varphi$. Now

$$\varphi_i \varphi_j = \varphi_i(\mathscr{E}_j s_j q\, \mu_j) = \mathscr{E}_i \mathscr{E}_j s_j q\, \mu_j = 0.$$

Using this, one can check that $\operatorname{Ker} \varphi = \bigoplus_{i \in I} X_i \cap \operatorname{Ker} \varphi_i$. □

In the following lemma, we list a few facts about summable families in S, which will be needed in the proof of the main theorem.

<u>**Lemma 3.23**</u>. (1) *If* $(g_j)_{j \in J}$ *and* $(f_i)_{i \in I}$ *are both summable, then so is* $(g_j f_i)_{J \times I}$ (*and consequently the order of summation* $\sum_j \sum_i g_j f_i$ *can be interchanged*).

(2) *If* $(g_i)_{i \in I}$ *is summable, and* $(f_i)_{i \in I}$ *is finitely valued* (in the sense that $\{f_i(m) : i \in I\}$ is finite for each $m \in M$), *then* $(g_i f_i)_{i \in I}$ *is summable.*

(3) *If* $(g_i)_{i \in I}$ *and* $(f_i)_{i \in I}$ *are both summable and* $g_i \equiv f_i$ (modulo Δ) *for all* $i \in I$, *then* $\sum g_i \equiv \sum f_i$.

PROOF. For $m \in M$, let $F(m) = \{i : f_i(m) \neq 0\}$ and $G(m) = \{j : g_j(m) \neq 0\}$.

(1) If $g_j f_i(m) \neq 0$, then $f_i(m) \neq 0$ and hence $i \in F(m)$, as well as $j \in G(f_i(m))$. Since $G(f_i(m)) \subseteq \bigcup_{k \in F(m)} G(f_k(m))$, which is finite, $g_j f_i$ is summable.

(2) Let $\{f_i(m) : i \in I\} = \{u_1, \ldots, u_t\}$, $u_i \in M$. If $g_i f_i(m) \neq 0$, then

$i \in G\,(f_i(m)) \subseteq \bigcup_{k=1}^{t} G(u_k)$, which is finite. Thus $(g_i f_i)_{i \in I}$ is summable.

(3) Without loss of generality we may assume $g_i = 0$, i.e., $f_i \in \Delta$. Consider any $0 \neq m \in M$. Then $\bigcap_{i \in F(m)} \operatorname{Ker} f_i \leq^e M$, hence the intersection contains $0 \neq mr$ for suitable $r \in R$. As $f_i(m) = 0$ for all $i \notin F(m)$, we obtain $mr \in \bigcap_{i \in I} \operatorname{Ker} f_i$. This proves that $\bigcap_{i \in I} \operatorname{Ker} f_i \leq^e M$. Since $\sum f_i$ vanishes on this submodule, $\sum f_i \in \Delta$. □

Now we are ready to prove the main theorem of this section.

<u>**Theorem 3.24**</u>. *Every continuous module has the exchange property.*

PROOF. Using Theorem 2.37, Lemma 3.20, and the fact that quasi–injective modules have the exchange property (Theorem 1.21), it suffices to establish the exchange property for a square free continuous module M. Here we know that all idempotents of $\overline{S} = S/\Delta$ are central (Lemma 3.4), $J = \Delta$ and $\overline{S}$ is regular (Proposition 3.5).

We establish the result by verifying (3) of Proposition 3.22. Let I be a set of ordinals, and $f_i \in S$ ($i \in I$) be a summable family with $\Sigma f_i = 1$. Since $\bar{S}$ is regular, there exist $\alpha_i \in S$ such that $f_i \equiv f_i \alpha_i f_i$ (modulo Δ). Let $h_i = \alpha_i f_i$; then clearly $(h_i)_{i \in I}$ is a summable family and the $\bar{h}_i$ are central idempotents in $\bar{S}$.

Inductively, we define $\mathscr{E}_k = (1 - \sum_{i<k} \mathscr{E}_i) h_k \in Sf_k$. By induction, we see that the $\mathscr{E}_k$ are well defined, summable, and are orthogonal idempotents modulo Δ. By Corollary 3.9, the $\mathscr{E}_k$ lift to orthogonal idempotents $g_k \in S$. Now

$$h_k = \mathscr{E}_k + (\sum_{i<k} \mathscr{E}_i) h_k \equiv \mathscr{E}_k + h_k \sum_{i<k} \mathscr{E}_i.$$

Then

$$\begin{aligned} 1 &= \sum_k f_k \equiv \sum_k f_k h_k \quad \text{(by (1) and (3) of Lemma 3.23)} \\ &\equiv \sum_k f_k (\mathscr{E}_k + h_k \sum_{i<k} \mathscr{E}_i) \quad \text{(by (2) of Lemma 3.23)} \\ &\equiv \sum_k (f_k \mathscr{E}_k + f_k \sum_{i<k} \mathscr{E}_i) \\ &= \sum_k \sum_{i \leq k} f_k \mathscr{E}_i \\ &= \sum_i \sum_{k \geq i} f_k \mathscr{E}_i \quad \text{(by (1) of Lemma 3.23).} \end{aligned}$$

Let $\varphi_i = \sum_{k \geq i} f_k$. Then

$$1 \equiv \sum_i \varphi_i \mathscr{E}_i \equiv \sum_i g_i \varphi_i \mathscr{E}_i.$$

Thus $\Sigma g_i \varphi_i \mathscr{E}_i = 1 + x$ for some $x \in \Delta$, so

$$\sum_i (1 + x)^{-1} g_i \varphi_i \mathscr{E}_i = 1 = \sum_i g_i \varphi_i \mathscr{E}_i (1 + x)^{-1}.$$

So $M = \oplus_i g_i M$, and

$$M = (1 + x)^{-1} M = \oplus_i (1 + x)^{-1} g_i M.$$

Let $(e_i)_{i \in I}$ be the natural projections of M with respect to the decomposition $M = \oplus (1 + x)^{-1} g_i M$. Since and $\sum_i (1 + x)^{-1} g_i \varphi_i \mathscr{E}_i = 1$, we get $e_i = (1 + x)^{-1} g_i \varphi_i \mathscr{E}_i \in Sf_i$ for all $i \in I$. Thus (3) of Proposition 3.22 holds and M has the exchange property. □

<u>Corollary 3.25</u>. *Directly finite continuous modules have the cancellation property.*

PROOF. A directly finite (quasi–)continuous module has the internal cancellation property (Theorem 2.33). The result then follows from the previous theorem and Proposition 1.23. □

Quasi–continuous modules, in general, do not have the exchange property, nor the cancellation property.

<u>Examples 3.26</u>. (1) A quasi–continuous directly finite module which fails to have the cancellation property: Swan [62] (Theorem 3) gives the well known example of a commutative domain R with a stably free projective module P which is not free, in fact $P \oplus R \cong R^{n+1}$ but $P \ncong R^n$. Thus R_R does not have the cancellation property. Since R_R is uniform, it is quasi–continuous and directly finite.

(2) A quasi–continuous directly finite module which fails to have the (finite) exchange property, but still has the cancellation property: The ring $\mathbb{Z}$ of integers (Fuchs [72], p. 210).

(3) A non–continuous, quasi–continuous directly finite module which has the exchange property, and therefore the cancellation property: Any local commutative domain R which is not a field (cf. Warfield [72]).

COMMENTS

Proposition 3.5 was first proved for injective modules by Utumi [59], and later generalized to quasi–injective modules by Faith and Utumi [64]. For an injective module M, Wong and Johnson [59] proved that if $\Delta = 0$, then S is right self–injective. This result was generalized by Osofsky [68c], who proved that if M is quasi–injective, then S/Δ is right self–injective (Theorem 3.10) and orthogonal idempotents modulo Δ lift (cf. Corollary 3.9).

For a right continuous ring R, Utumi [65] proved $J(R) = \Delta(R)$ and that R/Δ is right continuous. Mohamed and Bouhy [77] generalized this result to continuous modules (Proposition 3.5 and Theorem 3.11). The endomorphism ring of a quasi–continuous module was studied by Jeremy [74]; most of his results are stated without proof. Proposition 3.12 and Corollary 3.13 are due to the authors. Their application to right quasi–continuous right non–singular rings (Remark (2)) generalizes the decomposition obtained for right continuous regular rings by Utumi [60]; cf. Goodearl [79], Theorem 13.17). A related decomposition for arbitrary right continuous rings is discussed in Birkenmeier [76].

Most of the material in Section 2 is taken from Müller and Rizvi [83]. Theorem 3.17 was first proved for quasi–injective modules by Bumby [65]. Theorem 3.16 is due to Mucke [88]; the special case of indecomposable M_α's appears in Müller and Rizvi [84].

Proposition 3.22 is due to Zimmermann–Huisgen and Zimmermann [84]; the special case of the finite exchange property goes back to Nicholson [77]. Theorem 3.24 is new, cf. Mohamed and Müller [88b]. The finite exchange property of continuous modules was known before and follows from Warfield [72] and the information concerning the endomorphism ring in Section 1. Corollary 3.25 and the Examples 3.26 are taken from Müller and Rizvi [83].

CHAPTER 4

QUASI DISCRETE MODULES

In this, and the following chapters, we will study modules with properties that are dual to continuity and quasi–continuity. Such modules will be called discrete and quasi–discrete, respectively. This terminology, which is new, is chosen because these modules decompose into direct sums of indecomposables: a quasi–discrete module M has a decomposition, unique up to isomorphism, $M = \oplus H_i$, where the H_i are hollow; moreover if M is discrete, then the H_i have local endomorphism rings.

In several cases we determine when, conversely, a direct sum of hollow modules is quasi–discrete. In full generality this question is open. In contrast to the fact that every quasi–injective module is continuous, projective modules need not be quasi–discrete; we shall see that this implication remains intact if and only if projective covers exist.

1. DEFINITIONS AND BASIC RESULTS

Dual to the notion of essential submodules, we have

Definitions 4.1. A submodule A of a module M is called *small* in M (notation $A << M$) if $A + B \neq M$ for any proper submodule B of M. A module H is called *hollow* if every proper submodule of H is small.

The sum of all small submodules of a module M is equal to the Jacobson radical of M, and will be denoted by Rad M. Thus an arbitrary sum of small submodules of M is small if and only if $\text{Rad}\, M << M$. A finite sum of small submodules of M is always small in M.

Examples of modules that are equal to their radicals are C_p^∞ and $\mathbb{Q}$ as modules over $\mathbb{Z}$; the first is hollow while the second one is not.

There are two types of hollow modules H:

(i) $H \neq \text{Rad}\, H$; in this case $H = xR$ holds for every $x \notin \text{Rad}\, H$. Such modules, for obvious reasons, will be called *local* modules. These modules are dual to uniform modules with non–zero socle.

(ii) $H = \text{Rad}\, H$; such modules are dual to uniform modules with zero socle. It is obvious that they are not finitely generated, and may have a complicated structure, even over commutative noetherian rings (see Chapter 5).

The following lemma contains some facts about small submodules which will be used freely.

<u>Lemma 4.2</u>. *Let* A, B *and* C *be submodules of* M:

(1) *If* $A << B$ *and* $B \leq C$, *then* $A << C$;

(2) *If* $A << M$, $A \leq B$ *and* $B \subset^{\oplus} M$, *then* $A << B$;

(3) *If* $A << M$ *and* $\varphi : M \longrightarrow N$ *is a homomorphism, then* $\varphi A << \varphi M$.

PROOF. (1) Let $A + D = C$. Then $A + B \cap D = B$, and hence $B \cap D = B$ since $A << B$. This implies $B \leq D$ and so $A \leq D$. Thus $D = C$.

(2) Write $M = B \oplus B^*$, and let $A + D = B$. Then $A + D + B^* = M$. Since $A << M$, $D \oplus B^* = M$, consequently $D = B$.

(3) We may assume that φ is onto. Let $\phi A + D = N$. Then

$$M = \phi^{-1}N = \phi^{-1}\phi A + \phi^{-1}D = A + \phi^{-1}D.$$

Since $A << M$, $M = \phi^{-1}D$, and so

$$N = \phi M = \phi\phi^{-1}D \leq D.$$

Hence $D = N$. □

<u>Definition 4.3</u>. Let A and P be submodules of M. P is called a *supplement* of A if it is minimal with the property $A + P = M$. L is called a *supplement submodule* if L is a supplement of some submodule of M.

Complement submodules exist by Zorn's Lemma; in fact if $A, B \leq M$ with $A \cap B = 0$, then B is contained in a complement of A. However, supplement submodules need not exist, e.g. no non–trivial submodule of $\mathbb{Z}_{\mathbb{Z}}$ has a supplement.

<u>Definition 4.4</u>. A module M is called *supplemented* if for any two submodules A and B with $A + B = M$, B contains a supplement of A.

The following lemma provides a criterion to check when a submodule is a supplement.

<u>Lemma 4.5</u>. *Let* A *and* P *be submodules of* M. *Then* P *is a supplement of* A *if and only if* $M = A + P$ *and* $A \cap P << P$.

PROOF. Assume that P is a supplement of A, and let $P = A \cap P + D$. Then

$$M = A + P = A + A \cap P + D = A + D.$$

Minimality of P then implies $D = P$. Hence $A \cap P << P$.

Conversely, assume the condition, and let $M = A + Q$ with $Q \leq P$. Then $P = Q + A \cap P$, and since $A \cap P << P$, $P = Q$. Therefore P is a supplement of A.

□

Recall conditions (C_1), (C_2) and (C_3), and note that (C_1) is equivalent to the following condition:

(C_1') For every submodule A of M, there is a decomposition $M = M_1 \oplus M_2$ such that $A \leq M_1$ and $A + M_2 \leq^e M$.

The conditions (C_i) dualize as follows, respectively:

(D_1) For every submodule A of M, there is a decomposition $M = M_1 \oplus M_2$ such that $M_1 \leq A$ and $A \cap M_2 << M$;

(D_2) If $A \leq M$ such that M/A is isomorphic to a summand of M, then A is a summand of M.

(D_3) If M_1 and M_2 are summands of M with $M_1 + M_2 = M$, then $M_1 \cap M_2$ is a summand of M.

First we investigate some basic properties of the conditions (D_i).

<u>Lemma 4.6</u>. *Let* M *be a module with* (D_2). *Then*

(i) *if* $M_1, M_2 \subset^{\oplus} M$, *then any epimorphism* $M_1 \xrightarrow{f}\!\!> M_2$ *splits; and*

(ii) M *has* (D_3).

PROOF. (i) Write $M = M_1 \oplus M_1^*$. Then

$$M_2 \cong M_1/\text{Ker } f \cong (M_1 \oplus M_1^*)/(\text{Ker } f \oplus M_1^*) = M/(\text{Ker } f \oplus M_1^*).$$

Consequently $\text{Ker } f \oplus M_1^*$, and hence $\text{Ker } f$ is a summand of M_1.

(ii) Let $A, B \subset^{\oplus} M$ with $A + B = M$, and write $M = A \oplus A^*$. Then

$$A^* \cong (A + B)/A \cong B/A \cap B.$$

Hence $A \cap B \subset^{\oplus} M$ by (i).

□

<u>Lemma 4.7</u>. *Any summand of a module* M *with* (D_i) *also satisfies* (D_i).

PROOF. Straightforward.

□

Proposition 4.8. *The following are equivalent for a module* M:

(1) M *has* (D_1);

(2) *Every submodule* A *of* M *can be written as* $A = N \oplus S$ *with* $N \subset^{\oplus} M$ *and* $S << M$;

(3) M *is supplemented and every supplement submodule of* M *is a summand.*

PROOF. (1) ⇒ (2): M has a decomposition $M = M_1 \oplus M_2$ with $M_1 \leq A$ and $A \cap M_2 << M$. Then $A = M_1 \oplus A \cap M_2$, and the result follows with $N = M_1$ and $S = A \cap M_2$.

(2) ⇒ (3): Let $M = X + Y$; we show that Y contains a supplement of X. By assumption, we may assume $Y \subset^{\oplus} M$. Now $X \cap Y = Y_1 \oplus S$ such that $Y_1 \subset^{\oplus} M$ and $S << M$. Since $Y \subset^{\oplus} M$, $S << Y$. Write $Y = Y_1 \oplus Y_2$, and let π denote the projection $Y_1 \oplus Y_2 \longrightarrow\!\!> Y_2$. Then $X \cap Y = Y_1 \oplus X \cap Y \cap Y_2$, and

$$X \cap Y_2 = X \cap Y \cap Y_2 = \pi(X \cap Y) = \pi(Y_1 + S) = \pi S.$$

Hence $X \cap Y_2 << Y_2$. Now

$$M = X + Y = X + Y_1 + Y_2 = X + Y_2,$$

so Y_2 is a supplement of X.

Now let P be a supplement submodule of M. Then there exists $K \leq M$ such that P is minimal with the property $K + P = M$. Since $P = L \oplus T$ with $L \subset^{\oplus} M$ and $T << M$, $M = K + L$. Then minimality of P implies $P = L$.

(3) ⇒ (1): Let $A \leq M$. Then A has a supplement B and B has a supplement M_1 such that $M_1 \leq A$ and $M_1 \subset^{\oplus} M$. Write $M = M_1 \oplus M_2$. Then

$$A = M_1 \oplus A \cap M_2$$

Also $M = M_1 + B$ and so

$$A = M_1 + A \cap B.$$

Let ν denote the projection $M_1 \oplus M_2 \longrightarrow\!\!> M_2$. Then

$$A \cap M_2 = \pi A = \pi(A \cap B)$$

Since B is a supplement of A, $A \cap B << M$ and hence $A \cap M_2 << M$. Thus M has (D_1). □

Corollary 4.9. *An indecomposable module* M *has* (D_1) *if and only if* M *is hollow.* □

<u>Definition 4.10</u>. A module M is called *discrete* if it has (D_1) and (D_2); M is called *quasi–discrete* if it has (D_1) and (D_3).

It is clear that every hollow module is quasi–discrete. Summands of (quasi–) discrete modules are(quasi–) discrete (Lemma 4.7).

The following is dual to the corresponding result for quasi–continuous modules (see Theorem 2.8).

<u>Proposition 4.11</u>. *A module* M *is quasi–discrete if and only if* M *is supplemented, and* $M = X \oplus Y$ *for any two submodules* X *and* Y *of* M *which are supplements of each other.*

PROOF. "Only if": By Proposition 4.8, M is supplemented and $X, Y \subset^{\oplus} M$. Hence $X \cap Y \subset^{\oplus} M$ by (D_3). Since $X \cap Y << M$, $X \cap Y = 0$.

"If": M has (D_1) by Proposition 4.8. Let $A, B \subset^{\oplus} M$ such that $M = A + B$. Now B contains a supplement B' of A which is a summand of M. Thus $A \cap B' << B' \leq M$, consequently $A \cap B' << A$. Thus A is a supplment of B', hence $M = A \oplus B'$ by hypothesis. Then $B = B' \oplus A \cap B$, so $A \cap B \subset^{\oplus} B$. □

2. DECOMPOSITION THEOREMS

We show that any quasi–discrete module M is a direct sum of hollows, and that these summands can be arranged to obtain a decomposition $M = M_1 \oplus M_2$ such that M_1 has small radical and M_2 is equal to its radical.

<u>Lemma 4.12</u>. *Let* A *be a submodule of a quasi–discrete module* M, *and* B *a supplement of* A. *If* C *is a supplement submodule of* M *contained in* A, *then* $C \cap B = 0$ *and* $C \oplus B \subset^{\oplus} M$.

PROOF. B and C are summands of M (Proposition 4.8). Writing $M = C^* \oplus C$ we get $A = A \cap C^* \oplus C$, hence $M = A \cap C^* + C + B$. By Proposition 4.8, $A \cap C^*$ contains a supplement D of $C + B$, which is a summand. Since $D \leq C^*$, $D \oplus C \subset^{\oplus} M$. Now $M = (D \oplus C) + B$, a sum of two summands, so $(D \oplus C) \cap B \subset^{\oplus} M$ by (D_3). However

$$(D \oplus C) \cap B \leq A \cap B << M.$$

Hence $(D \oplus C) \cap B = 0$, and $M = D \oplus C \oplus B$. □

<u>Corollary 4.13</u>. *In a quasi–discrete module* M, *the union of any chain of summands is a summand, and local summands are summands.*

PROOF. Let $\{A_\alpha\}$ be a chain of summands of M, and let $A = \bigcup_\alpha A_\alpha$. Let B be a supplement of A. Then $A_\alpha \cap B = 0$ for every α by Lemma 4.12, and hence $A \cap B = 0$, so $M = A \oplus B$. The last statement follows by Lemma 2.16. □

<u>Proposition 4.14</u>. *Let* N *be a summand and* H *a hollow summand of a quasi–discrete module* M. *Then either* $N + H \subset^{\oplus} M$ *and* $N \cap H = 0$, *or* $N + H = N \oplus S$ *with* $S << M$ *and* H *is isomorphic to a summand of* N.

PROOF. Let L be a supplement of N + H. Then $N \cap L = 0$ by Lemma 4.12. Now $M = N + L + H$. We consider the two possible cases:

(i) $H \nleq N \oplus L$: Then H is a supplement of N + L, hence $N \cap H = 0$ and $N \oplus H \subset^{\oplus} M$ by Lemma 4.12.

(ii) $H \leq N \oplus L$: Then $M = N \oplus L$ and

$$N + H = N \oplus (N + H) \cap L = N \oplus S,$$

where $S = (N + H) \cap L << M$ since L is a supplement of N + H. Write $M = H \oplus H'$. Then

$$M = N + H + H' = N + S + H' = N + H'.$$

Let N′ be a supplement of H′ contained in N. Then $M = N' \oplus H'$ by Proposition 4.11. Hence $H \cong N' \subset^{\oplus} N$. □

We are now ready to prove the main theorem of this section.

<u>Theorem 4.15</u>. *Any quasi–discrete module* M *has a decomposition* $M = \bigoplus_{i \in I} H_i$ *where each* H_i *is hollow. Moreover, such a decomposition complements summands, and hence is unique up to isomorphism.*

PROOF. By Theorem 2.17 and Corollary 4.13, $M = \bigoplus_{i \in I} H_i$ where the H_i are indecomposable. It follows by Corollary 4.9 that each H_i is hollow.

Next we prove that the decomposition $M = \bigoplus_{i \in I} H_i$ complements summands. Let X be any summand of M. By Zorn's Lemma and Corollary 4.13, we get a subset $J \subseteq I$ maximal such that $X \cap (\bigoplus_{j \in J} H_j) = 0$, and $X \oplus (\bigoplus_{j \in J} H_j) \subset^{\oplus} M$. Write $M = X \oplus (\bigoplus_{j \in J} H_j) \oplus Y$. The proof will be complete if we show that Y = 0. Assume that $Y \neq 0$. Since Y is quasi–discrete, Y contains a non–zero hollow summand H (by what we have already proved). Consequently $M = T \oplus H$ where $X \oplus (\underset{j \in J}{\oplus} H_j) \leq T$. If for some $i \in I$, $T + H_i \subset^{\oplus} M$, then $T \cap H_i = 0$, and we get a contradiction to the maximality of J. Thus $T + H_i$ is not a summand of M for any $i \in I$. It then follows by Proposition 4.14 that for all $i \in I$

$$T + H_i = T \oplus S_i \ ; \ S_i << M$$

Consider any finite subset $F \subseteq I$. Then $T + (\underset{i \in F}{\oplus} H_i) \neq M$, and

$$M = T \oplus H = T + (\underset{i \in F}{\oplus} H_i) + H.$$

Hence H is a supplement of $T + (\underset{i \in F}{\oplus} H_i)$. But then $H \cap (\underset{i \in F}{\oplus} H_i) = 0$ by Lemma 4.12. Since this is true for any finite subset $F \subseteq I$, $H = H \cap (\underset{i \in I}{\oplus} H_i) = 0$, a contradiction. Therefore $Y = 0$, and $M = X \oplus (\underset{j \in J}{\oplus} H_j)$. □

The following corollary is a consequence of Theorems 2.26 and 4.15.

Corollary 4.16. *Any decomposition of a quasi–discrete module as a direct sum of hollows is locally semi–*T*–nilpotent.* □

The next proposition characterizes quasi–discrete modules with small radical.

Proposition 4.17. *The following are equivalent for a quasi–discrete module* M:

(1) Rad $M << M$;

(2) *Every proper submodule of* M *is contained in a maximal submodule;*

(3) M *is a direct sum of local modules.*

PROOF. (1) ⇒ (3): By Theorem 4.15, $M = \underset{i \in I}{\oplus} H_i$ with each H_i hollow. Since Rad $H_i \leq$ Rad $M << M$, Rad $H_i \neq H_i$. Hence H_i is local for all $i \in I$.

(3) ⇒ (2): Let A be a proper submodule of M. By Proposition 4.8, $A = N \oplus S$ where $N \subset^{\oplus} M$ and $S << M$. Now M has a decomposition $M = \underset{j \in J}{\oplus} M_j$ with each M_j local. It then follows by Theorem 4.15, that this decomposition complements summands. Hence $M = N \oplus (\underset{k \in K}{\oplus} M_k)$, for some subset $K \subseteq J$. For a fixed $\alpha \in K$, it is clear that

$$B = N \oplus (\bigoplus_{k \neq \alpha} M_k) \oplus \text{Rad } M_\alpha$$

is a maximal submodule of M. Now

$$A = N \oplus S \leq N + \text{Rad } M \leq B.$$

Hence (2) follows.

(2) ⇒ (1): Trivial. □

Corollary 4.18. *A quasi–discrete module* M *has a decomposition, unique up to isomorphism,* $M = M_1 \oplus M_2$, *where* M_1 *has small radical and* M_2 *is equal to its radical.*

PROOF. By Theorem 4.15, $M = \bigoplus_{i \in I} H_i$ with each H_i hollow. Define $J = \{j \in I : \text{Rad } H_j \neq H_j\}$ and let $K = I - J$. Let

$$M_1 = \bigoplus_{j \in J} H_j \ , \ M_2 = \bigoplus_{k \in K} H_k.$$

Since H_j is local for all $j \in J$, M_1 has small radical by Proposition 4.17. Now

$$\text{Rad } M_2 = \bigoplus_{k \in K} \text{Rad } H_k = \bigoplus_{k \in K} H_k = M_2.$$

The uniqueness of the decomposition $M = M_1 \oplus M_2$ follows from the uniqueness of the original decomposition $M = \bigoplus_{i \in I} H_i$. □

3. APPLICATIONS OF THE DECOMPOSITION THEOREMS

This section contains results which are dual to those in Section 4 of Chapter 2. The proofs are greatly facilitated by the availability of the decompositon theorems.

Corollary 4.19. *A quasi–discrete module* M *has the exchange property if and only if every hollow summand of* M *has a local endomorphism ring.*

PROOF. By Theorem 4.15, $M = \bigoplus_{i \in I} H_i$ where each H_i is hollow and the decomposition complements summands. The result now follows by Theorem 2.25 and the fact that an indecomposable module has the exchange property if and only if it has a local endomorphism ring (Warfield [69a]). □

Corollary 4.20. *A directly finite quasi–discrete module* M *has the internal cancellation property; and* M *has the cancellation property if every hollow summand of* M *has a local endomorphism ring.*

PROOF. The first statement is a consequence of Theorems 2.27 and 4.15. The second statement then follows from Proposition 1.23 and Corollary 4.19. □

Remark. The requirement that the endomorphism ring of each hollow summand is local, is automatically satisfied if M is discrete (see Corollary 5.5). Thus a discrete module has the exchange property, and a directly finite discrete module has the cancellation property.

In the following we discuss some properties of quasi–discrete modules which are analogous to those given in Lemma 2.30, Theorem 2.31 and Corollary 2.32.

Proposition 4.21. *Let* A *be a submodule of a quasi–discrete module* M. *If* P_1 *and* P_2 *are supplements of* A, *then* P_1 *and* P_2 *are perspective*; *in particular* $P_1 \cong P_2$.

PROOF. Let X be a supplement of P_1 contained in A. Then $M = X \oplus P_1$ by Proposition 4.11. Hence $A = X \oplus A \cap P_1$. Since $A \cap P_1 << M$,

$$M = A + P_2 = X + A \cap P_1 + P_2 = X + P_2.$$

It follows by Lemma 4.12 that $X \cap P_2 = 0$; and so $M = X \oplus P_2$. □

Lemma 4.22. *Let* M *be a quasi–discrete module. If* $M = \sum_{i \in I} M_i$ *is an irredundant sum of indecomposable submodules* M_i, *then* $M = \bigoplus_{i \in I} M_i$.

PROOF. The irredundancy of the sum $\sum_{i \in I} M_i$ implies that no M_i is small in M. Hence M_i is a hollow summand of M for every $i \in I$, by Proposition 4.8 and Corollary 4.9.

Let F be a finite subset of I and let K be a subset of F maximal such that $\sum_{i \in K} M_i$ is direct and a summand of M. Assume that $K \neq F$, and let $j \in F - K$. By maximality of F and Proposition 4.14

$$\bigoplus_{i \in K} M_i + M_j = \bigoplus_{i \in K} M_i \oplus S$$

where $S << M$. However, this implies $M = \sum_{i \neq j} M_i$, which is a contradiction to the irredundancy of the sum. Therefore $K = F$, and consequently $\sum_{i \in I} M_i$ is direct.

□

Lemma 4.23. *If* $A \oplus B$ *is quasi-discrete and* $X \leq B$, *then any homomorphism* $\phi : A \longrightarrow B/X$ *lifts to a homomorphism* $\psi : A \longrightarrow B$. *In particular, any epimorphism* $B \longrightarrow\!\!> A$ (or $A \longrightarrow\!\!> B$) *splits.*

PROOF. Let π denote the natural homomorphism $B \longrightarrow\!\!> B/X$ and define $\eta : M = A \oplus B \longrightarrow B/X$ by

$$\eta(a + b) = \phi(a) + \pi(b).$$

Let $K = \operatorname{Ker} \eta$. For an arbitrary element $a \in A$, $\phi(a) = \pi(b)$ for some $b \in B$, and hence $a - b \in K$. Therefore $A \leq K + B$, and consequently $M = K + B$. Let C be a supplement of B contained in K. Then $M = C \oplus B$ by Proposition 4.11. Let p denote the projection $C \oplus B \longrightarrow\!\!> B$, and $\psi = p|_A$. Then for every $a \in A$,

$$\phi(a) = \eta(a) = \eta((1-p)(a) + p(a)) = \eta(p(a) = \pi(p(a)) = \pi\ \psi(a).$$

Hence $\pi\psi = \phi$. The last statement is obvious. □

Theorem 4.24. *Let* A *and* B *be summands of a quasi-discrete module* M. *If* $A/X \cong B/Y$ *where* $X << A$ *and* $Y << B$, *then* $A \cong B$.

PROOF. Since A is quasi-discrete, $A = \bigoplus_{i \in I} A_i$ where each A_i is hollow (Theorem 4.15). Let $\overline{A} = A/X$ and $\overline{B} = B/Y$. Let θ be an isomorphism of $\overline{A}$ onto $\overline{B}$. Then $\overline{B} = \sum_{i \in I} \theta\overline{A}_i$. Since θ is an isomorphism and $X << A$, $\sum_{i \in I} \theta\overline{A}_i$ is irredundant. Let C_i be the full inverse image of $\theta\overline{A}_i$ in B; then it is clear that $B = \sum_{i \in I} C_i$ and that the sum is irredundant.

By Proposition 4.8, $C_i = B_i \oplus S_i$ where $B_i \subset^{\oplus} B$ and $S_i << B$. Since $\overline{C}_i \cong \overline{A}_i$, $\overline{C}_i$ is hollow, and hence $\overline{C}_i = \overline{B}_i$ or $\overline{C}_i = \overline{S}_i$. However $\overline{C}_i = \overline{S}_i$ would imply $C_i = S_i + Y << B$, which is a contradiction to the irredundancy of the sum $\sum_{i \in I} C_i$. Thus $\overline{C}_i = \overline{B}_i$ and hence $C_i = B_i + Y$.

Then

$$B = \sum_{i \in I} C_i = \sum_{i \in I} (B_i + Y) = \sum_{i \in I} B_i + Y$$

Since $Y << B$, $B = \sum_{i \in I} B_i$; and it is clear that the sum is irredundant.

Next we show that B_i is hollow. Assume that $B_i = L + N$. Then $\overline{B}_i = \overline{L} + \overline{N}$. As $\overline{B}_i$ is hollow, $\overline{B}_i = \overline{L}$ or $\overline{B}_i = \overline{N}$. Let us assume that $\overline{B}_i = \overline{L}$. Then $B_i + Y = L + Y$. Write $B = B_i \oplus D_i$. Then $B = B_i \oplus D_i = B_i + Y + D_i = L + Y + D_i = L \oplus D_i$, since $Y << B$. This implies that $B_i = L$. Similarly $\overline{B}_i = \overline{N}$ implies $B_i = N$. Thus B_i is hollow.

It now follows by Lemma 4.22 that $B = \bigoplus_{i \in I} B_i$. Since A_i and B_i are hollow summands of M, it follows by Proposition 4.14 that $A_i \cong B_i$ or $A_i + B_i$ is direct and is a summand of M. In the latter case $A_i \oplus B_i$ is quasi–discrete. Then by Lemma 4.23, there exists a homomorphism $\varphi_i : A_i \longrightarrow B_i$ such that the following diagram is commutative

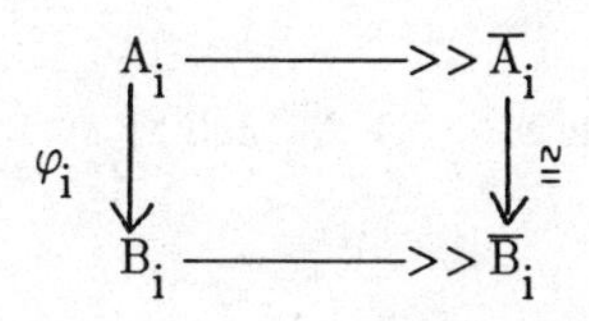

Since B_i is hollow, φ_i is onto. Then φ_i splits by Lemma 4.23, consequently φ_i is an isomorphism since A_i is hollow. Thus one has $A_i \cong B_i$ in either case. Hence

$$A = \bigoplus_{i \in I} A_i \cong \bigoplus_{i \in I} B_i = B \qquad \square$$

The following immediate consequence should be compared with Corollary 2.32.

Corollary 4.25. *Let* M *be a quasi–discrete module. Let* N_1 *and* N_2 *be submodules of* M, *with supplements* P_1 *and* P_2 *respectively. If* $M/N_1 \cong M/N_2$, *then* $P_1 \cong P_2$.

□

Definition 4.26. A module M is said to have the *lifting property* if for any index set I and any submodule X of M, if $M/X = \bigoplus_{i \in I} A_i$, then there exists a decomposition $M = M_0 \oplus (\bigoplus_{i \in I} M_i)$ such that (i) $M_0 \leq X$, (ii) $\overline{M}_i = A_i$, and (iii) $X \cap (\bigoplus_{i \in I} M_i) << M$.

If M has this property with respect to small submodules X of M, we say that M has the *small lifting property*. For n ∈ ℕ, the n–*lifting property* (n–*small lifting property*) will mean that M has the lifting property (small lifting property) for index sets with cardinality n.

If X is a small submodule of M, it is clear that $M_0 = 0$ and that (iii) above is automatically satisfied. Thus the small lifting property amounts to the requirement that every decomposition of M/X can be realized by a decomposition of M. It is also clear that the 1–lifting property is equivalent to (D_1). In the following we prove that the 2–lifting property is equivalent to the (full) lifting property.

Lemma 4.27. *Let* M *be a module with* (D_1). *Then the following are equivalent:*

(1) M *has* (D_3),

(2) *If for summands* M_1 *and* M_2 *of* M, $M = M_1 + M_2$ *and* $M_1 \cap M_2 << M$, *then* $M_1 \cap M_2 = 0$,

(3) *If for summand* $M_i (i \in I)$ *of* M, $M = \sum_{i \in I} M_i$ *and* $M_j \cap \sum_{i \neq j} M_i << M$ *for every* j, *then* $\sum_{i \in I} M_i$ *is direct.*

PROOF. (3) implies (2) trivially. We show that (2) implies (1) and (1) implies (3).

Assume (2), and let $A, B \subset^{\oplus} M$ be such that $M = A + B$. By Proposition 4.8, B contains a supplement N of A; and N is a summand of M. Now $M = A + N$ and $A \cap N << M$ imply $A \cap N = 0$ by hypothesis, so $M = A \oplus N$. Then

$$B = B \cap M = B \cap (A \oplus N) = A \cap B \oplus N,$$

hence $A \cap B \subset^{\oplus} M$. Thus M has (D_3) and (1) follows.

Assume (1), and let $M = \sum_{i \in I} M_i$ with the given conditions, and without loss of generality assume that each $M_i \neq 0$. We first note that the sum $\sum_{i \in I} M_i$ is irredundant. Indeed, if $M = \sum_{i \neq j} M_i$ for some $j \in I$, then $M_j = M_j \cap M = M_j \cap \sum_{i \neq j} M_i << M$, consequently $M_j = 0$, a contradiction. Next we observe that M_j is a supplement of $\sum_{i \neq j} M_i$ for every $j \in J$, since $M_j \cap \sum_{i \neq j} M_i << M_j$ by Lemma 4.2.

Now consider a finite subset $F \subseteq I$, and let $K \subseteq F$ be maximal such that $\sum_{i \in K} M_i$ is direct and is a summand of M. We claim that $K = F$. Suppose not; and let $j \in F - K$. Since M has (D_1) and (D_3), M is quasi–discrete. Since M_j is a supplement of $\sum_{i \neq j} M_i$ and $\sum_{i \in K} M_i \leq \sum_{i \neq j} M_i$, it follows by Lemma 4.12 that $\sum_{i \in K} M_i + M_j$ is direct and is a summand of M, in contradiction to the maximality of consequently $\sum_{i \in I} M_i$ is direct. □

Theorem 4.28. *The following are equivalent for a module* M:

(1) M *is quasi-discrete,*

(2) M *has the lifting property,*

(3) M *has the 1-lifting property and the 2-small lifting property.*

PROOF. (2) implies (3) trivially. We show that (3) implies (1) and (1) implies (2).

Assume (3). The 1-lifting property implies that M has (D_1). Now let $A_1, A_2 \subset^{\oplus} M$ such that $M = A_1 + A_2$ and $A_1 \cap A_2 << M$. Write $X = A_1 \cap A_2$ and $\overline{M} = M/X$. Then $\overline{M} = \overline{A}_1 \oplus \overline{A}_2$, and it follows by the 2-small lifting property that $M = M_1 \oplus M_2$ such that $\overline{A}_i = \overline{M}_i$ $(i = 1,2)$. Therefore $A_i = M_i + X$; and since $X << A_i$ by Lemma 4.2, $A_i = M_i$ $(i = 1,2)$. Hence $A_1 \cap A_2 = 0$. Then M has (D_3) by Lemma 4.27.

Assume (1). Let $X \leq M$ and $\overline{M} = M/X = \bigoplus_{i \in I} A_i$. By (D_1), $M = M_o \oplus N$ where $M_o \leq X$ and $X^* = X \cap N << M$. As $N \subset^{\oplus} M$, $X^* << N$. Now

$$N/X^* \cong (M_o \oplus N)/(M_o \oplus X^*) = M/X = \bigoplus_{i \in I} A_i.$$

Let N_i be the full inverse image of A_i in N. Again by Proposition 4.8 , $N_i = M_i \oplus S_i$ where $M_i \subset^{\oplus} N$ and $S_i << N$. Then $A_i = \overline{N}_i = \overline{M}_i + \overline{S}_i$. As $\overline{S}_i << \overline{N}$ by Lemma 4.2 and $A_i \subset^{\oplus} \overline{N}$, $\overline{S}_i << A_i$; consequently $A_i = \overline{M}_i$. Thus $\overline{N} = \bigoplus_{i \in I} \overline{M}_i$, which implies that $N = \sum_{i \in I} M_i + X^*$. Therefore $N = \sum_{i \in I} M_i$ since $X^* << N$. Since N is quasi-discrete and $M_j \cap \sum_{i \neq j} M_i \leq X^* << M$, $\sum_{i \in I} M_i$ is direct by Lemma 4.27. Hence $M = M_o \oplus (\bigoplus_{i \in I} M_i)$, and (2) follows. □

4. DISCRETENESS AND PROJECTIVITY

We start by summarizing properties of relative projectivity which are analogous to those in Section 1 of Chapter 1. We then discuss when (quasi-)projective modules are discrete. We finally investigate when quasi-discreteness passes down from a cover.

Definition 4.29. A module N is said to be *A-projective* if for every submodule X of A, any homomorphism $\varphi : N \longrightarrow A/X$ can be lifted to a homomorphism $\psi : N \longrightarrow A$.

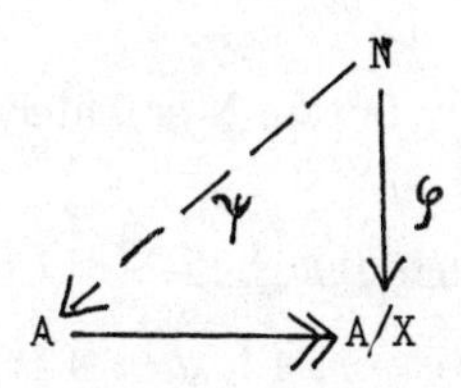

A module P is *projective* if P is A–projective for every R–module A. A module M is called *quasi–projective* if M is M–projective.

The following immediate consequence will be used frequently:

Lemma 4.30. *If* N *is* A*–projective, then any epimorphism* $A \xrightarrow{f} N$ *splits. If, in addition,* A *is indecomposable, then* f *is an isomorphism.* □

The following results concerning A–projectivity will be stated without proofs. The proofs are either dual to, or can be obtained by similar arguments as the proofs of the corresponding results in Chapter 1.

Proposition 4.31. *Let* N *be* A*–projective. If* $B \leq A$*, then* N *is* B*–projective and* A/B*–projective.* □

Proposition 4.32. *A direct sum* $\bigoplus_{\alpha \in \Lambda} M_\alpha$ *is* A*–projective if and only if* M_α *is* A*–projective for every* $\alpha \in \Lambda$. □

Proposition 4.33. *A module* N *is* $(\bigoplus_{i=1}^{n} A_i)$*–projective* ($n \in \mathbb{N}$) *if and only if* N *is* A_i*–projective,* $i = 1,2,\ldots,n$. □

The conclusion of Proposition 4.33 does not extend to infinite direct sums, as is shown by the following:

Example 4.34. Let I be an infinite set, and let $A_i = \mathbb{Z}$ for every $i \in I$. Trivially $\mathbb{Q}$ is A_i–projective. Obviously there exists an epimorphism $\bigoplus_{i \in I} A_i \longrightarrow \mathbb{Q}$. It does not split since $\mathbb{Q}_{\mathbb{Z}}$ is not projective. Hence $\mathbb{Q}$ is not $(\bigoplus_{i \in I} A_i)$–projective by Lemma 4.30.

Note that it also follows by Proposition 4.31 that $\mathbb{Q}$ is not $(\prod_{i \in I} A_i)$–projective.

In case N is finitely generated, Proposition 4.33 extends to infinite direct sums.

Proposition 4.35. *Let* I *be an arbitrary set. If* N *is finitely generated and* A_i*–projective for every* $i \in I$*, then* N *is* $(\bigoplus_{i \in I} A_i)$*–projective.*

PROOF. Let $X \leq \underset{i\in I}{\oplus} A_i$. Then $(\underset{i\in I}{\oplus} A_i)/X = \underset{i\in I}{\Sigma} \overline{A}_i$ where $\overline{A}_i = (A_i + X)/X$. For any homomorphism $\varphi : N \longrightarrow \underset{i\in I}{\Sigma} \overline{A}_i$, $\operatorname{Im} \varphi \leq \underset{i\in F}{\Sigma} \overline{A}_i$ for some finite $F \subseteq I$. Then Proposition 4.33 applies to the finite direct sum $\underset{i\in F}{\oplus} A_i$ and the result follows.

□

We also have the following corollaries to Propositions 4.32 and 4.33.

Corollary 4.36. *A direct sum of projective modules is projective.* □

Corollary 4.37. *A finite direct sum* $\overset{n}{\underset{i=1}{\oplus}} M_i$ *is quasi–projective if and only if* M_i *is* M_j*–projective* (i, j = 1, 2, ..., n). M^n *is quasi–projective if and only if* M *is quasi–projective.* □

Recall the hierarchy:

Injective ⇒ quasi–injective ⇒ continuous ⇒ quasi–continuous ⇒ (C_1).

In the present situation, we have

Projective ⇒ quasi–projective ⇏ discrete ⇒ quasi–discrete ⇒ (D_1).

In fact, a projective module need not have (D_1); e.g. $\mathbb{Z}_{\mathbb{Z}}$.

Proposition 4.38. *Any quasi–projective module* M *has* (D_2).

PROOF. Let $M \overset{f}{\longrightarrow\!\!\!\!\to} M'$ be an epimorphism with $M' \subset^{\oplus} M$. Then M' is M–projective by Proposition 4.32; hence f splits by Lemma 4.30. □

Now we characterize quasi–projective modules which are discrete.

Proposition 4.39. *A quasi–projective module* M *is discrete if and only if every submodule of* M *has a supplement.*

PROOF. If M is discrete, then it has (D_1); hence M is supplemented by Proposition 4.8.

Conversely, assume that every submodule of M has a supplement. We first show that M is supplemented. (Note that, in general, if every submodule of a module N has a supplement, N need not be supplemented; cf. Appendix).

Let $M = A + B$. We show that B contains a supplement of A. By assumption A has a supplement P. Then $M = A + P$ and $A \cap P << P$. Let ν and π be the natural homomorphisms $M \longrightarrow\!\!> M/A$ and $B \longrightarrow\!\!> M/A$, respectively. Since M is B–projective by Proposition 4.31, there exists $f : M \longrightarrow B$ such that $\pi f = \nu$. Let $\mu = \nu|_P$ and $g = f|_P$. Then $\pi gP = \mu P = M/A$, and hence $M = A + gP$. It is easy to check that $A \cap gP = g(\ker \mu)$. Since $\ker \mu = A \cap P << P$, $g(\ker \mu) << gP$ by Lemma 4.2. Hence $A \cap gP << gP$ and consequently gP is a supplement of A contained in B.

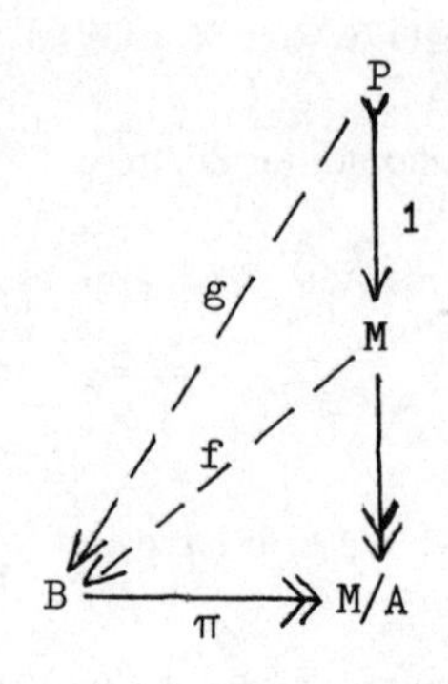

Next we prove that every supplement submodule of M is a summand; then it follows by Proposition 4.8 that M has (D_1). Suppose that A is a supplement submodule of M and B is a supplement of A, and use the same diagram. Since $\pi fA = \nu A = 0$, $fA \leq A$. Then

$$M = fM + A = f(A + B) + A = fA + fB + A = fB + A.$$

Then minimality of B implies $fB = B$, hence $M = B + \ker f$. Since $\ker f \leq A$, the minimality of A implies $\ker f = A$. Therefore

$$\ker f = A = \ker \nu = \ker \pi f.$$

Hence π is a monomorphism on $fM = B$; consequently $A \cap B = 0$. Hence $M = A \oplus B$.

Since M has (D_2) by Proposition 4.38, M is discrete. □

Every module is a homomorphic image of a free (hence projective) module. An epimorphism $P \xrightarrow{\eta}\!\!> M$ with P projective, is called a *projective cover* of M if $\ker \eta << P$. The notion of a projective cover is dual to that of an injective hull. However, projective covers need not exist; for instance $\mathbb{Q}_{\mathbb{Z}}$ does not possess a projective cover. If a module M has a projective cover P, then P is unique up to isomorphism.

A ring R is *right* (*semi*) *perfect* if every (finitely generated) R–module has a projective cover. These rings were introduced by Bass [60], and were studied by many authors. For a detailed survey of these rings we refer the reader to Faith [76a] and Anderson and Fuller [73].

In particular, a ring R is right semiperfect if and only if R/RadR is semisimple and idempotents modulo RadR can be lifted. Thus right semiperfect rings are left semiperfect and vice versa. Also we note that for a ring R to be semiperfect it suffices that every simple R–module has a projective cover.

<u>Lemma 4.40</u>. *Let* M = A + B. *If* M/A *has a projective cover, then* B *contains a supplement of* A.

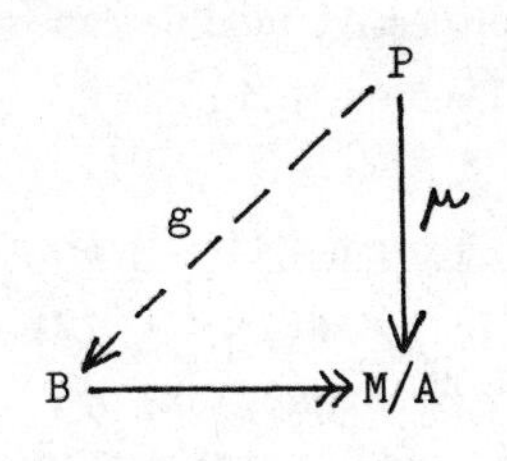

PROOF. Let $P \xrightarrow{\mu}\!\!> M/A$ be a projective over. Let π denote the natural homomorphism $B \longrightarrow\!\!>\!\!> M/A$. Since P is projective, there exists a homomorphism $g : P \longrightarrow B$ such that $\pi g = \mu$. As in the proof of Proposition 4.39, we get that gP is a supplement of A contained in B.

□

<u>Theorem 4.41</u>. *The following are equivalent for a ring* R:

(1) R *is a right (semi) perfect,*

(2) *Every (finitely generated) quasi–projective* R*–module is discrete,*

(3) *Every (finitely generated)* R*–module is supplemented,*

(4) *Every (cyclic) free* R*–module has the property that every submodule has a supplement.*

PROOF. (1) ⇒ (3) by Lemma 4.40, (3) ⇒ (2) by Proposition 4.39, and (2) ⇒ (4) trivially.

Assume (4), and let M be a (cyclic) R–module. Then there exists an epimorphism $F \xrightarrow{\eta}\!\!>\!\!> M$ where F is a (cyclic) free module. Since F has (D_1) by Proposition 4.39, $F = F_1 \oplus F_2$ with $F_1 \leq \ker \eta$ and $F_2 \cap \ker \eta << F_2$. Then it is clear that $\eta|_{F_2} : F_2 \longrightarrow\!\!>\!\!> M$ is a projective cover of M. Hence R is right (semi) perfect.

□

<u>Corollary 4.42</u>. *A ring* R *is semiperfect, if and only if* R_R *is discrete, if and only if every right ideal of* R *has a supplement.* □

Mares [63] defined a module P to be *semiperfect* if P is projective and every homomorphic image of P has a projective cover. The following is an immediate consequence of the definition and the proof of Theorem 4.41.

<u>Corollary 4.43</u>. *A projective module* P *is semiperfect, if and only if* P *is discrete, if and only if every submodule of* P *has a supplement.* □

We now give a characterization of semiperfect modules which is analogous to that of a semiperfect ring; this characterization shows that the discreteness of a projective module can be characterized by a weaker version of the lifting property (see Theorem 4.28).

<u>**Theorem 4.44**</u>. *A projective module* P *is semiperfect if and only if* (1) $\mathrm{Rad}P \ll P$, (2) $P/\mathrm{Rad}P$ *is semisimple, and* (3) *decompositions of* $P/\mathrm{Rad}P$ *lift to decompositions of* P.

PROOF. "If" : We first show that if P is any module with (1), (2) and (3), then every submodule of P has a supplement. Let A be a submodule of P. Then by (2) $\overline{P} = \overline{A} \oplus \overline{B}$ where $B \leq P$ ($\overline{P} = P/\mathrm{Rad}P$), and by (3) $P = P_1 \oplus P_2$ such that $\overline{P}_1 = \overline{A}$ and $\overline{P}_2 = \overline{B}$. Then

$$P = P_1 \oplus P_2 = P_1 + \mathrm{Rad}P + P_2 = A + \mathrm{Rad}P + P_2,$$

and by (1) we get $P = A + P_2$. Now

$$A \cap P_2 \leq A \cap (P_2 + \mathrm{Rad}P) = A \cap (B + \mathrm{Rad}P) \leq \mathrm{Rad}P.$$

Hence $A \cap P_2 \ll P$ by (1); thus P_2 is a supplement of A. (In fact we have proved that every submodule of P has a supplement which is a summand). It now follows by 4.43 that a projective module with (1), (2) and (3) is semiperfect.

Conversely, assume that P is semiperfect. Then P is discrete by Corollary 4.43. Using Corollary 4.18 and the fact that a non–zero projective module is not equal to its radical, we get $\mathrm{Rad}P \ll P$. Thus (1) holds.

Next we show that (2) holds for any module P whose submodules have supplements. Indeed, let K be a submodule of $\overline{P}$. Then $K = \overline{A}$ for some submodule $A \leq P$. Let B be a supplement of A. Then $P = A + B$ with $A \cap B \ll B$. Thus $A \cap B \ll P$ and hence $A \cap P \leq \mathrm{Rad}P$. Therefore $\overline{P} = \overline{A} \oplus \overline{B}$, and (2) holds.

Finally (3) holds by Theorem 4.28. □

Corollary 1.14 may be dualized as follows: If M has a projective cover $P \xrightarrow{\eta} M$, then M is quasi–projective if and only if ker η is invariant under every endomorphism of P (see Wu and Jans [67]); in fact the same conclusion holds if P is only a quasi–projective cover.

A dual statement to the equivalence of (1) and (3) in Theorem 2.8 is not true in general, even in case M has a projective cover; this is due to the fact that projective modules need not be discrete. However we have the following analogous result.

Proposition 4.45. *Let* N *be quasi–discrete; and let* $N \xrightarrow{f} \twoheadrightarrow M$ *be an epimorphism with a small kernel. Then* M *is quasi–discrete if and only if* ker f *is invariant under every idempotent of* End N.

PROOF. Assume that M is quasi–discrete and let e be an idempotent of End N. Write $N = A_1 \oplus A_2$ where $A_1 = eN$ and $A_2 = (1-e)N$. Then $M = fA_1 + fA_2$, and by quasi–discreteness, $M = B_1 \oplus B_2$ with $B_i \leq fA_i$ $(i = 1,2)$. Hence $N = f^{-1}B_1 + f^{-1}B_2$. Now $f^{-1}B_i = A_i \cap f^{-1}B_i + \ker f$ $(i = 1,2)$. Since $\ker f << N$,

$$N = A_1 \oplus A_2 = A_1 \cap f^{-1}B_1 \oplus A_2 \cap f^{-1}B_2.$$

Thus $A_i = A_i \cap f^{-1}B_i$, and so $A_i \leq f^{-1}B_i$, and consequently $fA_i = B_i$ $(i = 1,2)$. Hence $M = fA_1 \oplus fA_2$, which implies that $e \ker f \leq \ker f$. (Note that this direction of the proof does not require N to be quasi–discrete!)

Conversely, assume that ker f is invariant under every idempotent of EndN. Let A be an arbitrary submodule of M. Since N has (D_1),

$$N = X \oplus Y \text{ with } X \leq f^{-1}A \text{ and } Y \cap f^{-1}A << N.$$

Then $f^{-1}A = X \oplus S$ where $S = Y \cap f^{-1}A$. Since f is onto,

$$A = ff^{-1}A = fX + fS.$$

Now the hypothesis on kerf yields $fX \subset^{\oplus} M$ and $A = fX \oplus fS$. Since $fS << M$ by Lemma 4.2, M has (D_1).

Note that if $A \subset^{\oplus} M$, then $fS << A$, so $A = fX$. This shows that every summand of M is the image of some summand of N. Now let B_1 and B_2 be summands of M such that $M = B_1 + B_2$. Then there exists $C_i \subset^{\oplus} N$ such that $fC_i = B_i$ $(i = 1,2)$. Thus

$$N = C_1 + C_2 + \ker f = C_1 + C_2$$

since $\ker f << N$. Since N has (D_3), $C_1 \cap C_2 \subset^{\oplus} M$. Hence

$$N = C_1' \oplus C_1 \cap C_2 \oplus C_2'$$

where $C_i = C_i' \oplus C_1 \cap C_2$ $(i = 1,2)$. The hypothesis on kerf yields

$$M = fC_1' \oplus f(C_1 \cap C_2) \oplus f(C_2').$$

Consequently $B_1 \cap B_2 = f(C_1 \cap C_2) \subset^{\oplus} M$. Hence M has (D_3). □

A quasi–projective module which has a projective cover, need not be projective. The following example shows that a similar observation holds for a quasi–discrete module with a discrete cover.

Example 4.46. Let R be a discrete valuation ring and let K be its quotient field. As R–modules, K is indecomposable and discrete, K/R is hollow (hence quasi–discrete) but not discrete, and K ——>> K/R is a cover.

5. QUASI–DISCRETENESS OF DIRECT SUMS

By Theorem 4.15, every quasi–discrete module is a direct sum of hollow modules. According to Lemma 4.23, these hollow modules are relatively projective. We shall investigate now when, conversely, a direct sum of relatively projective hollow modules is quasi–discrete.

Lemma 4.47. *Let* $M = S \oplus T = N + T$ *where* S *is* T*–projective. Then* $M = S' \oplus T$ *where* $S' \leq N$.

PROOF. The hypothesis gives the following commutative diagram

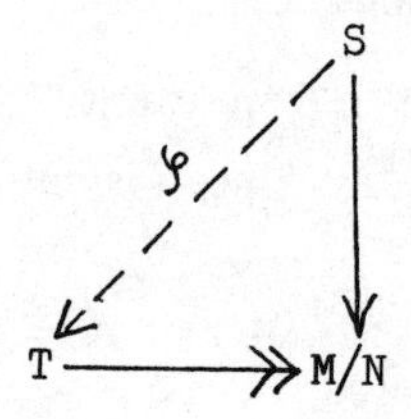

Let $S' = \{x - \varphi(x) : x \in S\}$. Then $S' \leq N$ and $M = S' \oplus T$. □

The conditions in the following theorem are strong, and hard to verify in concrete cases. However, the theorem allows a unified approach to the three subsequent corollaries, which represent the only general results known to–date. Moreover, by Harada's Theorem 2.25, (b) and (c) are each equivalent to the relatively easily verifiable condition lsTn, provided all M_i have local endomorphism rings.

Theorem 4.48. *Let* $M = \bigoplus_{i \in I} M_i$ *with* M_i *hollow. Then* M *is quasi–discrete if and only if:*

(a) M_i *is* $M(I-i)$*–projective for every* $i \in I$,

(b) $\bigoplus_{i \in I} M_i$ *complements summands, and*

(c) *every local summand of* M *is a summand.*

PROOF. The necessity of the conditions follows from 4.13, 4.15 and 4.23. Conversely assume the conditions. Then (a) in conjunction with Propositions 4.31 and 4.32 yields that M(J) and M(K) are relatively projective for any two disjoint subsets J and K of I. For any decomposition $M = N \oplus L$, (b) implies the existence of a subset J of I such that $N \cong M(J)$ and $L \cong M(I-J)$; consequently N and L are relatively projective. To derive (D_1), consider a submodule A of M. It follows from (c) and Lemma 2.16 that there exists a summand A^* of M maximal such that $A^* \leq A$. Write $M = A^* \oplus A^{**}$; then $A = A^* \oplus A \cap A^{**}$. Then (D_1) holds if we show that $B = A \cap A^{**} << M$.

Suppose, to the contrary, that B is not small in M. Then $M = B + C$ with $C \neq M$. As before we get a summand C^* of M maximal such that $C^* \leq C$. By (b), $M = C^* \oplus M(J)$ for some subset $J \subseteq I$. Since $C \neq M$, $J \neq \emptyset$. Consider $k \in J$ and let π denote the projection $C^* \oplus M(J-k) \oplus M_k \longrightarrow\!\!> M_k$. From $M = B + C$ we get $M_k = \pi B + \pi C$; and since M_k is hollow, $M_k = \pi B$ or $M_k = \pi C$. Thus $M = B + \ker \pi$ or $M = C + \ker \pi$. Lemma 4.47 then yields $M = D \oplus \ker \pi$ or $M = H \oplus \ker \pi$ where $D \leq B$, $H \leq C$ and $D \cong M_k \cong H$. In the first instance $A^* \oplus D$ contradicts the maximality of A^*, and in the second one $C^* \oplus H$ contradicts the maximality of C^*. This proves $B << M$, and (D_1) holds.

Now we prove (D_3). Consider summands N and T of M such that $M = N + T$. Write $M = S \oplus T$. Then S is T–projective. Applying Lemma 4.47 again we get $M = S' \oplus T$ with $S' \leq N$. It follows that $N = S' \oplus N \cap T$, and consequently $N \cap T \subset^{\oplus} M$. □

<u>Corollary 4.49</u>. *Let* $M = \bigoplus_{i \in I} M_i$ *such that* M_i *is hollow with local endomorphism ring. Then* M *is quasi–discrete if and only if* M_i *is* $M(I-i)$*–projective for every* $i \in I$, *and* lsTn *holds.*

PROOF. By Theorem 2.25, lsTn holds for $M = \bigoplus_{i \in I} M_i$, if and only if the decomposition complements summands, if and only if every local summand of M is a summand. The claim now follows immediately from Theorem 4.48. □

<u>Remark</u>. The two conditions in Corollary 4.49 are independent, even if one assumes that the M_i are relatively projective in pairs.

For example, let R be a commutative noetherian complete valuation domain, with quotient field K. Then $R^{(\mathbb{N})}$ is projective, but obviously does not satisfy lsTn. On the other hand $K^{(\mathbb{N})}$ trivially fulfills lsTn, and K is quasi–projective (Lemma 5.11), but is not $K^{(\mathbb{N})}$ – projective (cf. the proof of Lemma 5.12).

Corollary 4.50. *Let* $M = \bigoplus_{i=1}^{n} M_i$ *where* M_i *is hollow and* M_j*–projective whenever* $j \neq i$. *Then* M *is quasi–discrete.*

PROOF. We verify the conditions (a), (b) and (c) of Theorem 4.48.

(a) holds by Proposition 4.33.

We prove (b) by induction on n. It is obvious that (b) holds for n = 1. Assume that (b) holds for n–1. Consider a nonzero summand A of M; let $\pi_i : \bigoplus_{i=1}^{n} M_i \longrightarrow\!\!\!\to M_i$ be the natural projections. Since $A \leq \bigoplus_{i=1}^{n} \pi_i A$ and each M_i is hollow, $\pi_k A = M_k$ for some $k \in \{1,2,...,n\}$. Thus $M = A + \ker \pi_k$. Set $N = \ker \pi_k = \bigoplus_{i \neq k} M_i$. Then M_k is N–projective and M = A + N. Applying Lemma 4.47 we get $M = A^* \oplus N$ with $A^* \leq A$, consequently $A = A^* \oplus A \cap N$.

Now $A \cap N \subset^{\oplus} N = \bigoplus_{i \neq k} M_i$ implies by induction hypothesis that $N = A \cap N \oplus M(J)$ where $J \subseteq \{1, ..., k-1, k+1, ..., n\}$. Then

$$M = A^* \oplus N = A^* \oplus A \cap N \oplus M(J) = A \oplus M(J).$$

Hence (b) holds.

It then follows that any summand of M is isomorphic to M(K) for some subset $K \subseteq \{1,2,...,n\}$. Thus if $\{N_\alpha : \alpha \in \Lambda\}$ is a local summand of M, then Λ has at most n elements, hence $\bigoplus_{\alpha \in \Lambda} N_\alpha \subset^{\oplus} M$, and (c) follows. □

Corollary 4.51. *Let* $M = \bigoplus_{i \in I} M_i$ *where each* M_i *is local and* M_j*–projective whenever* $j \neq i$. *Then* M *is quasi–discrete if* RadM << M.

PROOF. We first note that M_i is M(I–i)–projective by Proposition 4.35. Then, applying Propositions 4.31 and 4.32, we get that M(J) is M(I–J)–projective for any subset $J \subseteq I$.

Consider an arbitrary submodule A of M. Since $\overline{M}$ = M/RadM is semisimple, $\overline{M} = \overline{A} \oplus \overline{M(J)}$ for some subset $J \subseteq I$. As RadM << M,

$$M = A + M(J) \quad \text{and} \quad A \cap M(J) << M.$$

(We note in passing that, if $B \leq A$, then there exists $K \supseteq J$ such that $M = B + M(K)$ and $B \cap M(K) << M$.) Now relative projectivity of $M(J)$ and $M(I - J)$ yields, by Lemma 4.47, that $M = A^* \oplus M(J)$ with $A^* \leq A$. Hence $A = A^* \oplus A \cap M(J)$. (In fact this shows already that (D_1) holds). Now if $A \subset^{\oplus} M$, then $A \cap M(J) << A$ and consequently $A = A^*$. Thus $M = A \oplus M(J)$, hence the decomposition complements summands.

Let $\{A_\alpha : \alpha \in \Lambda\}$ be a chain of summands of M and let $A = \bigcup_{\alpha \in \Lambda} A_\alpha$. Then $M = A + M(J)$ with $A \cap M(J) << M$. As noted earlier, for any $\alpha \in \Lambda$, there exists $J_\alpha \geq J$ with $M = A_\alpha + M(J_\alpha)$ and $A_\alpha \cap M(J_\alpha) << M$; therefore $M = A_\alpha \oplus M(J_\alpha)$. In particular $A_\alpha \cap M(J) = 0$, and consequently $A \cap M(J) = 0$. Hence $M = A \oplus M(J)$, and $A \subset^{\oplus} M$. It follows then by Lemma 2.16 that every local summand of M is a summand.

The corollary now follows by Theorem 4.48. □

We complete these considerations by showing that the condition "Rad M << M" of Corollary 4.51 can be replaced by a number of other conditions, notably again by "lsTn".

<u>Lemma 4.52</u>. *Let* $M = \bigoplus_{i \in I} M_i$ *with* M_i *local, and* M_j*–projective whenever* $j \neq i$. *Let* $M = \text{Rad } M + X$ *for some* $X \leq M$, *and* $m \in M_\alpha - X$ *for some* $\alpha \in I$. *Then there exists* $\alpha \neq \beta \in I$ *and a non–isomorphism* $f : M_\alpha \longrightarrow M_\beta$ *such that* $f(m) \notin X$.

PROOF. Since M_α is local, $M_\alpha = uR$ for some $u \in M_\alpha$. Write $u = z + x$, with $z \in \text{Rad}M$, $x \in X$. There exists a finite subset $F \subseteq I$ such that

$$z = \sum_{i \in F} z_i \ , \quad z_i \in \text{Rad } M_i \text{ and } x = \sum_{i \in F} x_i \ , \quad x_i \in M_i.$$

It is clear that $\alpha \in F$, $u = z_\alpha + x_\alpha$ and $z_i + x_i = 0$ $(i \neq \alpha)$. Then $x_i \in \text{Rad } M_i$ $(i \neq \alpha)$ and $x_\alpha \notin \text{Rad } M_\alpha$. Thus $M_\alpha = x_\alpha R$ and $x_\alpha + y = x$ where $y = \sum_{\alpha \neq i \in F} x_i$. Let $N = \bigoplus_{\alpha \neq i \in F} M_i$ and define $\nu : M_\alpha \longrightarrow N/yx_\alpha^0$ by $\nu(x_\alpha r) = \overline{yr}$. It is clear that ν is well defined.

By relative projectivity we get $\varphi : M_\alpha \longrightarrow N$ such that the diagram commutes. Therefore $\overline{\varphi(x_\alpha)} = \bar{y}$, and consequently $\varphi(x_\alpha) - y = ya$ for some $a \in x_\alpha^o$. Then

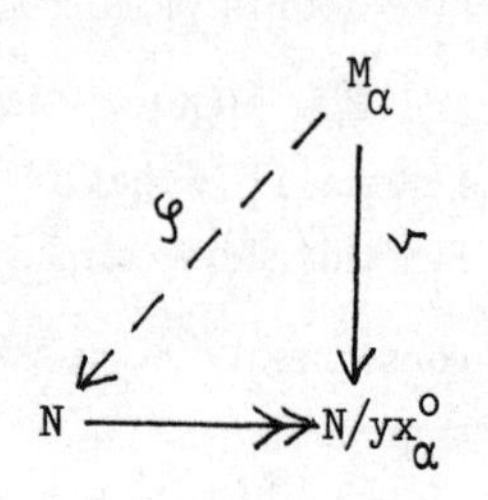

$$\varphi(x_\alpha) + x_\alpha = y(1 + a) + x_\alpha = y(1 + a) + x_\alpha(1 + a)$$
$$= (y + x_\alpha)(1 + a) = x(1 + a).$$

Let $m = x_\alpha r$. Then

$$\varphi(m) + m = x(1 + a)r \in X.$$

Since $m \notin X$ we conclude $\varphi(m) \notin X$. Let $\{\pi_i\}$ be the set of projections of N onto M_i $(i \in F - \{\alpha\})$. Then for some $\beta \in F - \{\alpha\}$, $\pi_\beta \varphi(m) \notin X$. Now

$$\pi_\beta\varphi(x_\alpha) = \pi_\beta(y(1 + a)) = \pi_\beta(\sum_{\alpha \neq i \in F} x_\alpha(1 + a)) = x_\beta(1 + a).$$

Consequently $\pi_\beta \varphi(x_\alpha) \in \text{Rad } M_\beta$. Thus $f = \pi_\beta\varphi : M_\alpha \longrightarrow M_\beta$ is the required non-isomorphism. □

Theorem 4.53. *Let* $M = \oplus_{i \in I} M_i$ *with* M_i *local and* M_j*–projective whenever* $j \neq i$. *Then the following are equivalent:*

(1) M *is quasi–discrete;*

(2) $\text{RadM} << M$;

(3) *Every proper submodule of* M *is contained in a maximal submodule;*

(4) *The decomposition* $M = \oplus_{i \in I} M_i$ *complements summands;*

(5) $\{M_i\}$ *is locally semi–T–nilpotent.*

PROOF. The equivalence of (1), (2) and (3) follows from Proposition 4.17 and Corollary 4.51.

We obtain (2) ⇒ (4) from Theorem 4.15, and (4) ⇒ (5) from Theorem 2.26. We proceed to prove (5) ⇒ (2). Assume, contrary to (2), that Rad M + X = M with $X \neq M$. Pick $\alpha \in I$ such that $M_\alpha \not\leq X$ and select $m \in M_\alpha - X$. We shall obtain a contradiction to lsTn, by constructing, inductively, a set of distinct indicies $\{\alpha_j : j \in \mathbb{N}\} \subseteq I$ and non–isomorphism $f_j : M_{\alpha_j} \longrightarrow M_{\alpha_{j+1}}$ such that $\alpha_1 = \alpha$ and $f_n \cdots f_2 f_1(m) \notin X$ for all $n \in \mathbb{N}$.

Assume that the construction is done for $j < n$. Let $I' = I - \{\alpha_1, \ldots, \alpha_{n-1}\}$, $M' = \oplus_{i \in I'} M_i$, $X' = X \cap M'$, and $m' = f_{n-1} \cdots f_1(m)$. Since $\text{RadM}_i << M_i \leq M$,

$$M = \mathrm{Rad}M + X = \mathrm{Rad}M_{\alpha_1} + \dots + \mathrm{Rad}\, M_{\alpha_{n-1}} + \mathrm{Rad}\, M' + X = \mathrm{Rad}M' + X.$$

Hence

$$M' = \mathrm{Rad}M' + X \cap M' = \mathrm{Rad}M' + X'.$$

Clearly $m' \in M_{\alpha_n} - X'$. Then Lemma 4.52 applies to M', and yields $\alpha_n \neq \alpha_{n+1} \in I'$ and a non–isomorphism $f_n : M_{\alpha_n} \longrightarrow M_{\alpha_{n+1}}$ such that $f_n(m') \notin X'$. Since $M_{\alpha_{n+1}} \cap X \leq M' \cap X = X'$, we have $f_n f_{n-1} \dots f_1(m) \notin X$.

□

Corollary 4.54. *A projective module is discrete if and only if it is a direct sum of local submodules and the radical is small (equivalently the decomposition is locally semi–T–nilpotent).* □

COMMENTS

Hollow modules were defined by Fleury [74a,b] and used by him, Rangaswamy [77] and Varadarajan [79a,b] in the context of dual Goldie dimension. Supplements appear first in Mares [63], and are studied, in their relationship with projective covers, by Kasch and Mares [66] and Miyashita [66]. Further investigations, in arbitrary modules and under various terminologies, are found in Zöschinger's papers (from [74a] onwards), Birjukov [78], Varadarajan [79a] and Hausen [82]. Discrete modules are defined, under the name dual–continuous, in Mohamed and Singh [77], and quasi–discrete ones, under the name quasi–semiperfect, in Oshiro [83a].

The decomposition theorem 4.15 is due to Oshiro [83a]. Preliminary versions appeared in Mohamed and Singh [77], Mohamed and Müller [79] and Kasch [79]; the present proof comes from Mohamed, Müller and Singh [85]. For projective modules, the theorem is already in Mares [63].

The exchange property (Corollary 4.19) follows immediately from the decomposition theorem and Zimmermann–Huisgen and Zimmermann [84]. There is an extensive literature concerning various aspects of the lifting property; cf. eg. Harada [80], [82a,c], Oshiro [83a], [84a,b].

The basic results on relative projectivity come from Azumaya [P] and Azumaya, Mbuntum and Varadarajan [75]. Proposition 4.39 is related to Golan [71b]. Theorem 4.41 was observed by several people, eg. Varadarajan [79a]. Theorem 4.44 is proved in Mares [63].

Most of the remaining results are taken from Mohamed and Müller [81], for discrete modules, and from Mohamed, Müller and Singh [85], for quasi–discrete modules. The projective case of Corollary 4.51 appears in Mares [63]. Lemma 4.52 and Theorem 4.53 are new.

CHAPTER 5

DISCRETE MODULES

The short first section investigates when a quasi–discrete module is discrete or quasi–projective. The next section studies the endomorphism ring of a (quasi–)discrete module, and contains results analogous to those in Section 1 of Chapter 3. The final section provides an explicit description of all discrete modules over commutative noetherian rings.

1. DISCRETE MODULES

The first lemma, and its proof, are dual to Lemma 3.14.

Lemma 5.1. *A quasi–discrete module* M *is discrete, if and only if every epimorphism* $M \longrightarrow\!\!> M$ *with small kernel is an isomorphism.*

PROOF. The necessity of the condition is obvious. Conversely, to establish (D_2), let $f : M \longrightarrow\!\!> N$ be an epimorphism, with kernel K, onto the summand N of M. As M is quasi–discrete, there is a decomposition $M = A \oplus B$ with $A \leq K$ and $B \cap K << B$. Now $N \cong M/K = K+B/K \cong B/B \cap K$. Theorem 4.24 yields $N \cong B$. Let $g : N >\!\!\longrightarrow\!\!> B$ be an isomorphism. Then $M = A \oplus B \xrightarrow{1 \oplus f|B} A \oplus N \xrightarrow{1 \oplus g} A \oplus B = M$ is an epimorphism with the small kernel $B \cap K$. By assumption, it is an isomorphism, that is $B \cap K = 0$. Then $K = A \oplus B \cap K = A \subset^{\oplus} M$, and f splits. □

The following theorem allows to determine when a quasi–discrete module is discrete or quasi–projective. Recall that a quasi–discrete quasi–projective module is always discrete (Proposition 4.39).

Theorem 5.2. *Let* $M = \bigoplus_{i \in I} M_i$ *be a quasi–discrete module, represented as a direct sum of hollow submodules, according to* Theorem 4.15. *Then*

(1) M *is discrete if and if all* M_i *are discrete;*

(2) M *is quasi–projective if and only if all* M_i *are quasi–projective.*

PROOF. The necessity of the conditions is obvious, as both properties are inherited by summands.

In the converse direction, in case (1), let $f : M \longrightarrow\!\!> M$ be an epimorphism with small kernel. Then $M = \Sigma\, fM_i$ is an irredundant sum of hollow submodules

fM_i. Lemma 4.22 implies $M = \underset{i \in I}{\oplus} fM_i$. By Theorem 4.15, there is a permutation p of I such that $fM_i \cong M_{p_i}$. The epimorphism $M_i \xrightarrow{f|M_i} fM_i \cong M_{p_i}$ is an isomorphism, since M_i is M_j–projective for $i \neq j$, and since M_i is discrete by assumption. It follows that f is an isomorphism. Hence M is discrete, by Lemma 5.1.

In case (2), M_i is $M(I-i)$–projective since M is quasi–discrete, and M_i–projective by assumption. Hence M_i is M–projective, and therefore M is M–projective (Propositions 4.32 and 4.33). □

2. ENDOMORPHISM RINGS

Throughout this section, S denotes the endomorphism ring of the module M, J the Jacobson radical of S, ∇ the collection of endomorphisms with small image, and $\bar{S} = S/\nabla$. (It is easy to see that ∇ is an ideal of S.)

Lemma 5.3. *If* M *is quasi–discrete, then idempotents modulo* ∇ *can be lifted.*

PROOF. Let $e \in S$ such that $e^2 - e \in \nabla$. Then $(e^2 - e)M$ is small. Since $M = eM + (1-e)M$ and M is quasi–discrete, there exists an idempotent $f \in S$ such that $fM \leq eM$ and $(1-f)M \leq (1-e)M$. Clearly $(f-e)M \leq eM$. Now, for any $m \in M$, $(f-e)m = (1-e)fm - e(1-f)m = (1-e)fm - e(1-e)m' = (1-e)(fm - em')$. Thus $(f-e)M \leq (1-e)M$. Hence $(f-e)M \leq eM \cap (1-e)M = (e-e^2)M << M$. We conclude $f - e \in \nabla$. □

The next theorem is analogous to Proposition 3.15.

Theorem 5.4. *Let* M *be quasi–discrete. Then* M *is discrete if and only if* $\nabla = J$ *and* S/∇ *is regular.*

PROOF. "Only if": Let α be an arbitrary element of S. Since M is (quasi–) discrete, $M = M_1 \oplus M_2$ with $M_1 \leq \alpha M$ and $\alpha M \cap M_2 << M$. Let e denote the projection $M_1 \oplus M_2 \longrightarrow\!\!> M_1$. Then $e\alpha : M \longrightarrow M_1$ is an epimorphism. Since M is discrete, $\text{Ker } e\alpha \subset^{\oplus} M$. Write $M = \text{Ker } e\alpha \oplus N$. The restriction of $e\alpha$ to N is an isomorphism onto M_1, and the inverse isomorphism of M_1 to N can be extended to an element $\beta \in S$. Then $\beta e \alpha = 1_N$. Now $(\alpha - \alpha\beta e\alpha)M = (\alpha - \alpha\beta e\alpha)(\text{Ker } e\alpha \oplus N) = \alpha(\text{Ker } e\alpha) \leq \alpha M \cap M_2$. Consequently $(\alpha - \alpha\beta e\alpha)M << M$, so $\alpha - \alpha\beta e\alpha \in \nabla$. Therefore S/∇ is a regular ring.

It then follows that $J \leq \nabla$. We proceed to show that $\nabla \subseteq J$. Given $\lambda \in \nabla$, $\lambda M << M$. Since $M = \lambda M + (1-\lambda)M$, $(1-\lambda)M = M$. As M is discrete, $1-\lambda$ is right invertible. But then $\lambda \in J$, since ∇ is an ideal of S.

"If" : Let $\varphi \in S$ be an epimorphism with small kernel. There exists $\psi \in S$ such that $\varphi - \varphi\psi\varphi \in \nabla$. We prove that $1 - \psi\phi \in \nabla$. Assume that $(1-\psi\varphi)M + B = M$. Then $\varphi(1-\psi\varphi)M + \varphi B = \varphi M = M$. Hence $\varphi B = M$ as $\varphi(1-\psi\varphi)M << M$. Thus $M = B + \ker\varphi = B$ since $\ker\varphi << M$. Hence$(1-\psi\varphi)M << M$. Then $1-\psi\varphi \in \nabla = J$, and hence $\psi\varphi$ is a unit in S. Thus φ is a monomorphism. Then M is discrete by Lemma 5.1. □

Corollary 5.5. *An indecomposable discrete module has local endomorphism ring.*
PROOF. S/∇ has no non–trivial idempotents by Lemma 5.3. It then follows by Theorem 5.4 that S/J is a division ring. □

Corollary 5.6. *Let* M *be a discrete module. Then* S *is semiperfect if and only if* M *is a direct sum of a finite number of hollow submodules.*
PROOF. By Theorem 4.15, M has a decomposition, unique up to isomorphism, $M = \bigoplus_{i \in I} H_i$ where each H_i is hollow. It also follows by Lemma 5.3 and Theorem 5.4 that any finite or countable set of orthogonal idempotents of S/J lifts to a set of orthogonal idempotents of S.

Assume that S is semiperfect. Then S contains a finite number of indecomposable orthogonal idempotents whose sum is 1. The uniqueness of the decomposition $M = \bigoplus_{i \in I} H_i$ then implies that I is finite.

Conversely, assume that I is finite. Again the uniqueness of the decomposition implies that S cannot contain more than $|I|$ orthogonal idempotents' the same is also true for S/J. Hence S/J is semisimple, consequently S is semiperfect. □

Theorem 3.11 says that S/Δ is a right continuous ring if M is a continuous module. The analogous statement for discrete M is not valid: S/∇ is right discrete, if and only if it is semisimple, if and only if M is a finite direct sum of hollow modules.

The next result should be compared with Corollary 3.13.

Proposition 5.7. *If* M *is quasi–discrete, then there is a ring decomposition* $\overline{S} = \overline{S}_1 \times \overline{S}_2$ *such that* $\overline{S}_1$ *is regular and* $\overline{S}_2$ *is reduced.*

PROOF. Let $M = A \oplus B$, where A and B have no nonzero isomorphic summands. We claim that fA is small, for every homomorphism $f : A \longrightarrow B$.

If not, fA contains a nonzero summand S of B. The induced epimorphism $A \longrightarrow\!\!> S$ splits, since S is A–projective. This produces a summand S′ of A isomorphic to S, in contradiction to the assumption.

We decompose $M = \bigoplus_{i \in I} H_i$ into hollow summands, according to Theorem 4.15. Let M_2 be the direct sum of those H_i which appear with multiplicity one, and M_1 the direct sum of the rest. By Proposition 5.2, M_1 is quasi–projective, hence discrete. By Theorem 5.4, $\overline{S}_1$ is regular.

The claim applies to $M = M_1 \oplus M_2$, and shows that fM_i is small, for any $f : M_i \longrightarrow M_j$ $(i \neq j)$. Consequently $\overline{S} = \overline{S}_1 \times \overline{S}_2$.

Let $\alpha \in S_2$ with $\alpha^2 M_2 << M_2$. Write $M_2 = A \oplus B$ with $A \leq \alpha M_2$ and $B \cap \alpha M_2$ small; let π_A and π_B be the corresponding projections. Then $\pi_B \alpha M_2 = B \cap \alpha M_2$ is small. Secondly $\pi_A \alpha A \leq \pi_A \alpha^2 M_2 << \pi_A M_2 = A$ is small. Thirdly $\pi_A \alpha B = \mathrm{im}(\pi_A \alpha|_B : B \longrightarrow A)$ is small by the claim. Consequently $\alpha M << M$, hence $\alpha \in \nabla_2$. This shows that $\overline{S}_2$ is reduced. □

The next lemma is partially an analogue of Lemma 3.2.

<u>Lemma 5.8</u>. *Let* M *be a quasi–discrete module. If* $\{e_i : i \in I\}$ *is a family of idempotents of* S *which are orthogonal modulo* ∇, *then* $\sum_{i \in I} e_i M$ *is direct, and is a summand of* M.

PROOF. Consider any finite subfamily $\{e_1, \dots, e_n\}$. As $(1 - \sum_{i=1}^{n} e_i)^2 = 1 - \sum_{i=1}^{n} e_i + \sum_{i \neq j} e_i e_j$ and $\sum_{i \neq j} e_i e_j \in \nabla$, the element $1 - \sum_{i=1}^{n} e_i$ of S is idempotent modulo ∇. By Lemma 5.3 it lifts to an idempotent $e_0 \in S$. It follows readily that the extended family $\{e_0, e_1, \dots, e_n\}$ is orthogonal modulo ∇, and sums to 1 modulo ∇.

We deduce immediately $M = \sum_{i=0}^{n} e_i M$. Since $e_j M \cap \sum_{i=0, i \neq j}^{n} e_i M \leq \sum_{i=0, i \neq j}^{n} e_j e_i M << M$, Lemma 4.27 implies $M = \bigoplus_{i=0}^{n} e_i M$.

We conclude that the sum $\sum_{i\in I} e_iM$ is direct, and is a local direct smmand of M. But then it is a summand of M, by Corollary 4.13. □

The last result of this section parallels Corollary 3.9.

Corollary 5.9. *If* M *is quasi–discrete, then any family of orthogonal idempotents of* $\overline{S}$ *lifts to a family of orthogonal idempotents of* S.

PROOF. Let $\{\bar{e}_i : i \in I\}$ be a family of orthogonal idempotents of $\overline{S}$. By Lemma 5.3 we may assume that each e_i is an idempotent of S. By Lemma 5.8 we have $M = \bigoplus_{i\in I} e_iM \oplus D$. It follows easily that $M = \bigoplus_{i\in I} e_iM \oplus (1-e_j)D$, for any $j \in I$. Let f_j be the projection onto e_jM corresponding to this last decomposition.

One has $f_j^2 = f_j$ and $e_jM = f_jM$, hence $e_jf_j = f_j$, $f_je_j = e_j$, and $f_je_i = 0$ for $j \neq i$. Consequently $f_if_j = f_ie_jf_j = 0$, and the f_j are orthogonal idempotents.

Moreover

$$(e_j - f_j)M = \sum_{i\in I} (e_j - f_j)e_iM + (e_j - f_j)(1-e_j)D = \sum_{i\in I,\, i\neq j} e_je_iM.$$

If the last sum is small, we can conclude $e_j - f_j \in \nabla$, and f_j is a lift for $\bar{e}_j$, as required.

Suppose, to the contrary, that $\sum_{i\in I, i\neq j} e_je_iM$ is not small in e_jM. Then it contains a nonzero hollow summand H of e_jM. Let π be the projection from e_jM to H. Then $\sum_{i\in I, i\neq j} \pi e_je_iM = H$, hence $\pi e_j : \bigoplus_{i\in I, i\neq j} e_iM \longrightarrow H$ is onto. As M is quasi–discrete, H is $(\bigoplus_{i\in I, i\neq j} e_iM)$ – projective. Therefore the epimorphism splits:

$$\bigoplus_{i\neq j} e_iM = A \oplus (\ker \pi e_j \cap \bigoplus_{i\neq j} e_jM),$$

with $A \cong H$. But $\bigoplus_{i\neq j} e_iM$ refines to a decomposition which complements summands (Theorem 4.15); hence

$$\bigoplus_{i\neq j} e_iM = B \oplus (\ker \pi e_j \cap \bigoplus_{i\neq j} e_iM)$$

where $B \subset^{\oplus} e_kM$ for some $k \neq j$. We apply πe_j and obtain $H = \pi e_j(\bigoplus_{i\neq j} e_iM) = \pi e_jB$. Since $e_je_kM << e_jM$, we conclude $\pi e_jB \leq \pi e_je_kM << \pi e_jM = H$. This yields the contradiction $H << H$. □

3. COMMUTATIVE NOETHERIAN RINGS

The aim of this section is to determine all discrete modules over commutative noetherian rings. The material is taken from Mohamed and Müller [88]. Because of the relative complexity of the arguments, we use freely results from Kaplansky [74], Matlis [73] and Zariski–Samuel [60].

R shall always be a commutative noetherian ring. Our starting point is an explicit description of the structure of hollow discrete modules.

Proposition 5.10. *Let* H *be a hollow module, over a commutative noetherian ring* R, *with annihilator* I.

H *is (discrete and) local, if and only if* R/I *is a local ring, and* H *is isomorphic to* R/I.

H *is discrete and non–local, if and only if* R/I *is a local ring with exactly one additional prime ideal,* P/I, *the integral closure of* R/P *is also local, and* H *is isomorphic to the total quotient ring* $(R/I)_P$ *of* R/I.

PROOF. Without loss of generality we assume I = 0. The statement concerning the local case is trivial.

Let H be hollow and discrete. We define $P = \{x \in R{:}Hx \neq H\}$; this is a prime ideal (in fact, it is the unique co–associated prime ideal of H; cf. Chambless [81], Zöschinger [83]). Clearly H is P–divisible. As the epimorphisms $x : H \longrightarrow H$ $(x \notin P)$ are isomorphisms, by (D_2), H is also P–torisonfree, and is therefore a module over the localization R_P. As such, it is still hollow and discrete.

The radical of H, as R_P–module, is $\mathrm{Rad}H = H\,\mathrm{Rad}R_P = HP = \sum_{i=1}^{n} Hp_i \neq H$, where $P = \sum_{i=1}^{n} p_iR$. Thus, as R_P–module, H is local, hence cyclic, hence isomorphic to R_P.

Observe that P is a maximal ideal, if and only if H is local as R–module: in the non–trivial direction, if P is maximal, the quotient field R_P/P_P of R/P coincides with R/P, and therefore $R_P = R + P_P$. As $H \cong R_P$ is hollow, we deduce $R_P = R$. Consequently, R and H are local.

Now we assume, in addition, that H is non–local. Thus P is non–maximal. According to Matlis [73], Theorem 10.5, the following are equivalent for a commutative noetherian domain which is not a field : (i) the domain is local and one–dimensional, and its completion has only one rank zero prime ideal; (ii) the integral closure is a discrete valuation ring; (iii) the quotient field is a hollow module.

As H is isomorphic to R_P, the quotient field R_P/P_P of R/P is still hollow as R/P–module, and Matlis' Theorem applies to R/P. We conclude that R/P is local and one–dimensional, with local integral closure.

To complete the proof in the non–local case, in one direction, it remains to show that R has no prime ideals beyond P and the unique maximal ideal $\mathfrak{m}$ containing P.

Suppose there is an additional prime ideal. If it is comparable with P, then it is properly contained in P, and we can choose it to be maximal such. Call it Q′. By Kaplansky [74], Theorem 144, there are infinitely many prime ideals between Q′ and $\mathfrak{m}$. Thus, in any case, we can find a prime ideal, Q, which is incomparable with P.

Let R_{PQ} denote the localization at the complement of $P \cup Q$, which has two maximal ideals, P_{PQ} and Q_{PQ}. One has $R \subset R_{PQ} \subset R_P$. Both R_{PQ}/P_{PQ} and R_P/P_P identify naturally with the quotient field of R/P, and hence with each other. One concludes $H \cong R_P = R_{PQ} + P_P$. Hollowness yields $R_P = R_{PQ}$; a contradiction.

In the converse direction, we have to show that any R_P, with all the properties of the non–local case, is hollow, non–local and discrete. As overrings of one–dimensional noetherian rings are noetherian and at most one–dimensional (Kaplansky [74], Theorem 93), the integral closure of R/P is a discrete valuation ring. Then Matlis' Theorem implies that R_P/P_P is hollow. But P is the prime radical of R, hence nilpotent. It follows that R_P itself is hollow.

Any R–homomorphism $R_P \longrightarrow R_P$ is automatically an R_P–homomorphism. Thus (D_2) holds, and R_P is discrete. R_P is non–local, as P is not maximal.

□

According to the proposition, for any hollow discrete module H, there is a unique maximal ideal containing the annihilator of H. We shall call it the *attached maximal ideal* of H.

From now on, we shall write any hollow discrete R–module as R/I or $(R/I)_P$, and tacitly assume the properties of the proposition. In the context of the next lemma, we note that hollow discrete modules with distinct attached maximal ideals are always relatively projective, since there are no non–zero homomorphisms between subfactors.

Lemma 5.11. *For hollow discrete modules with the same attached maximal ideal* $\mathfrak{m}$, *relative projectivity holds for*

(i) R/I_1 *and* R/I_2 *if and only if* $I_1 = I_2$;

(ii) R/I_1 *and* $(R/I_2)_P$ *if and only if* $I_1 \leq I_2$;

(iii) $(R/I_1)_{P_1}$ *and* $(R/I_2)_{P_2}$ *if and only if either* $P_1 \neq P_2$, *or* $P_1 = P_2$ *and* $I_1 = I_2$ *and* R/I_1 *is complete.*

PROOF. (i) Suppose that R/I_1 is R/I_2–projective. Consider the two natural maps $R/I_i \twoheadrightarrow R/(I_1 + I_2)$. One obtains $a \in R$ with $aI_1 \leq I_2$ and $1 - a \in I_1 + I_2$. Then $1 - a$ lies in the attached maximal ideal $\mathfrak{m}$. Hence $a \notin \mathfrak{m}$, and $\bar{a}$ is invertible in the local ring $R/(I_1 \cap I_2)$. One concludes $I_1 \leq I_2$. The rest is trivial.

(ii) Suppose that R/I_1 is $(R/I_2)_P$–projective. Consider the natural maps $R/I_i \longrightarrow R_P/(\bar{I}_1 + I_{2P})$. As before one obtains $a \in R_P$ with $aI_1 \leq I_{2P}$ and $1 - a \in \bar{I}_1 + I_{2P}$. If $I_1 \leq P$, then $1 - a \in P_P$ and hence $a^{-1} \in R_P$. Then $\bar{I}_1 \leq I_{2P} \cap \bar{R} = \bar{I}_2$, and therefore $I_1 \leq I_2$. If, on the other hand, $I_1 \nleq P$, then there is $b \in I_1 - P$, hence $b^{-1} \in R_P$. Thus $ab \in aI_1 \leq I_{2P}$ yields $a \in I_{2P}$, and consequently $1 = (1-a) + a \in \bar{I}_1 + I_{2P}$. This gives $1 = x + y$ with $x \in \bar{I}_1$ and $y = 1 - x \in I_{2P} \cap \bar{R} = \bar{I}_2$. There results $1 = x + y \in \bar{I}_1 + \bar{I}_2 \leq \bar{\mathfrak{m}}$; a contradiction.

In the converse direction, R/I_1 is $(R/I_2)_P$–projective since, due to $I_1 \leq I_2$, all R–modules to be considered are R/I_1–modules. $(R/I_2)_P$ is trivially R/I_1–projective, since all R–homomorphisms f from $(R/I_2)_P$ to factors $\bar{R}$ of R/I_1 are zero : indeed for $c \in \mathfrak{m} - P$ and arbitrary $x \in (R/I_2)_P$ and $n \in \mathbb{N}$, one has $f(x) = f(xc^{-n})c^n \in \bar{\mathfrak{m}}^n$, hence $f(x) \in \bigcap_{n=1}^{\infty} \bar{\mathfrak{m}}^n = 0$.

(iii) The first case is again trivial, since all maps are zero: given $f : (R/I_1)_{P_1} \longrightarrow \overline{(R/I_2)}_{P_2}$, we determine s such that $P_2^s \subset I_2$, and $d \in P_2^s - P_1$. Then for every $x \in (R/I_1)_{P_1}$, $f(x) = f(xd^{-1})d \in \overline{(R/I_2)}_{P_2} I_2 = 0$.

In the second case, $I_1 = I_2$ follows as in (i). We may then assume $I_1 = I_2 = 0$, and write $P_1 = P_2 = P$. To show that R is complete, we note first that the $\mathfrak{m}$–adic topology coincides with the cR–adic one, for any $c \in m - P$. Consider a sequence $v_n \in R$ with $v_{n+1} - v_n \in c^n R$. Then $f(c^{-n}) = v_n c^{-n} + R$ yields a well defined R–homomorphism $f : R_P \longrightarrow R_P/R$. By relative projectivity, there is $a \in R_P$ with $f(c^{-n}) = ac^{-n} + R$, that is $v_n - a \in c^n R$. We conclude $\lim v_n = a$, and so R is complete.

In the converse direction, we have to show that R_P is quasi–projective if R is complete. Consider any R–homomorphism $f: R_P \longrightarrow R_P/X$, where X is any R–submodule of R_P. Again, f is determined by $f(c^{-n}) = w_n + X$, with $w_n \in R_P$. The sequence $w_n c^n$ is Cauchy in the $\{Xc^n\}$–topology on R_P. We shall see that R_P is complete in this topology, so that $\lim w_n c^n = a$ exists in R_P. We obtain $w_n c^n - a \in Xc^n$ hence $w_n - ac^{-n} \in X$. It follows that a extends f.

To verify the completeness of R_P, we define $D = \bigcap_{n=1}^{\infty} Xc^n$. It obviously suffices to check the completeness of X/D. As D is c–divisible, it is an ideal of R_P, and is therefore of the form $D = (D \cap R)_P$. Passing to $R/D \cap R$ we may assume $D = 0$.

Define the ideals $A_n = Xc^n \cap R$ of R. Then $\bigcap_{n=1}^{\infty} A_n = 0$, $A_n c \leq A_{n+1} \leq A_n$, and the $A_n c^{-n}$ form an ascending sequence with union X. Since cR contains a power of the maximal ideal m, it follows from Chevalley's Theorem (cf. Zariski and Samuel [60], Theorem 13 on page 270) that there is N with $A_N \leq cR$. One deduces $A_n c = A_{n+1}$ for all $n \geq N$. Consequently $X = \bigcup_{n=1}^{\infty} A_n c^{-n} = A_N c^{-N} \cong A_N$ is finitely generated, and therefore complete. □

The next lemma discusses the discreteness of a direct sum of copies of the same hollow discrete module.

<u>Lemma 5.12</u>. *For a local hollow discrete module* R/I, *the direct sum* $(R/I)^{(n)}$ *is discrete if and only if* n *is finite or* R/I *is artinian. For a non–local hollow discrete module* $(R/I)_P$, $(R/I)_P^{(n)}$ *is discrete if and only if* n *is finite, and* R/I *is complete if* $n > 1$.

PROOF. We may again assume $I = 0$. By Theorems 4.53 and 5.2, $R^{(n)}$ is discrete if and only if it satisfies lsTn. This is true if n is finite. If n is infinite, since R is a local ring, lsTn holds precisely if RadR is T–nilpotent, ie. if R is a perfect ring. As R is noetherian, this means that it is artinian.

By Lemma 5.11, Corollary 4.50 and Theorem 5.2, the given conditions are certainly necessary and sufficient, for $(R/I)_P^{(n)}$ to be discrete, if n is finite. It remains to show that $(R/I)_P^{(\mathbb{N})}$ is not discrete.

Fix, as before, an element $c \in m-P$, and recall that the c^{-n} generate R_P as R–module. Thus

$$f(c^{-n}) = (c^{-n},..., c^{-1}, 0, ...) + R^{(\mathbb{N})}$$

determines a well defined R–homomorphism $R_P \longrightarrow R_P^{(\mathbb{N})}/R^{(\mathbb{N})}$. Suppose that $R_P^{(\mathbb{N})}$ is discete. Then R_P is $R_P^{(\mathbb{N})}$–projective (Lemma 4.23), and we obtain a lift $R_P \longrightarrow R_P^{(\mathbb{N})}$. As this R–homomorphism is automatically an R_P–homomorphism, we have an element $a = (a_1,...,a_N,0,...) \in R_P^{(\mathbb{N})}$ such that $(c^{-n},...,c^{-1},0,...) - (a_1,...,a_N,0,...)c^{-n} \in R^{(\mathbb{N})}$ for all n. For $n = N+1$, the $(N+1)$–coordinate yields $c^{-1} \in R$; a contradiction. □

We need an auxiliary general result.

<u>Lemma 5.13</u>. *Let* R *be a commutative noetherian ring. Let* I_j $(j \in \mathbb{N})$ *be a sequence of ideals, all contained in the same maximal ideal* m. *Then, after suitable reindexing, the following is true: for any choice of* s *and* N *there exists* t *such that* $\bigcap_{j=s}^{t} I_j \subset I_1 + m^N$.

PROOF. We claim first that, for almost all k, I_k contains $\bigcap_{j \notin F} I_j$ for all finite F.

Indeed, suppose there are infinitely many exceptions $k_1 < k_2 < ....$ This means that there are $x_n \in \bigcap_{j \notin F_n} I_j - I_{k_n}$ for all n. As R is noetherian, we have $\sum_{n=1}^{\infty} x_n R = \sum_{n=1}^{s} x_n R$. Choose q such that $k_q \notin \bigcup_{n=1}^{s} F_n$. Then $x_n \in I_{k_q}$ holds for $n = 1,...,s$. But then $x_q \in \sum_{n=1}^{s} x_n R \leq I_{k_q}$, which is a contradiction.

We apply this claim to the ideal $\hat{I}_j$ of the m–adic completion $\hat{R}$ of R. We reindex such that it is valid for k = 1, and obtain then in particular that $\bigcap_{j=s}^{\infty} \hat{I}_j \leq \hat{I}_1$ holds for all s.

Let $D = \bigcap_{j=s}^{\infty} \hat{I}_j$. Then $\hat{R}/D$ is noetherian, local and complete. By Chevalley's Theorem (Zariski and Samuel [60], VIII. 5.13), there exists $t = t(N)$ such that $\bigcap_{j=s}^{t} \hat{I}_j \leq \hat{m}^N + D$. We write $\overline{R} = R/\bigcap_{n=1}^{\infty} m^n$, and deduce $\bigcap_{j=s}^{t} \overline{I}_j \leq \bigcap_{j=s}^{t} \hat{I}_j \leq \hat{m}^N + D \leq \hat{m}^N + \hat{I}_1 = (m^N + I_1)\hat{}$ hence $\bigcap_{j=s}^{t} \overline{I}_j \leq (m^N + I_1)\hat{} \cap \overline{R} = (m^N + I_1)^-$, using that $m^N + I_1$ is co–artinian hence m–adically closed. The lemma follows immediately.

□

The crucial step towards the main theorem is the following observation.

Lemma 5.14. *An infinite direct sum of non–local hollow discrete modules with the same attached maximal ideal is never dual continuous.*

PROOF. Using the second part of Lemma 5.12, one sees that it suffices to derive a contradiction, from the assumption that a countable direct sum $\bigoplus_{k=0}^{\infty} (R/I_k)_{P_k}$ with distinct P_k is dual continuous.

We assume that the indexing is chosen in such a way that Lemma 5.13 applies. We observe that $(R/I_0)_{P_0}$ is $\bigoplus_{k=1}^{\infty} (R/I_k)_{P_k}$–projective. But $\hom_R((R/I_0)_{P_0}, \bigoplus_{k=1}^{\infty} (R/I_k)_{P_k}) = 0$ holds by the proof of Lemma 5.11 (iii). So we conclude that $\hom_R((R/I_0)_{P_0}, \bigoplus_{k=1}^{\infty} (R/I_k)_{P_k}/X) = 0$ holds for all R–submodules X.

We construct a specific homomorphism f. We start by fixing an element $c \in m - P_0$. Then $m^N + P_0 \leq c^n R + P_0$ holds for suitable $N = N(n)$. Lemma 5.13 shows that we can define, inductively, an increasing function g such that $g(0) = 1$ and $\bigcap_{j=g(n-1)+1}^{g(n)} I_j \leq I_0 + m^N \leq P_0 + c^n R$.

Next we put $w_n = \sum_{j=g(n-1)+1}^{g(n)} e_j$, where $e_j = (0,\ldots,0,1,0,\ldots) \in \bigoplus_{k=1}^{\infty} (R/I_k)_{P_k}$. We let $X = \Sigma_n (w_{n+1} c - w_n)R + \Sigma_n w_n I_0$, and determine $f : (R/I_0)_{P_0} \longrightarrow \bigoplus_{k=1}^{\infty} (R/I_k)_{P_k}/X$ by $f(c^{-n}) = w_n + X$. This yields a well defined R–homomorphism, which must be zero. In particular $f(c^{-1}) = 0$, and we obtain $r_n \in R$, $x_n \in I_0$ such that $w_1 = \Sigma_n(w_{n+1}c - w_n)r_n + \Sigma_n w_n x_n$, or $\Sigma_n w_n(\delta_{n1} - cr_{n-1} + r_n - x_n) = 0$. Since, by our choice of w_n, the sum $\Sigma_n w_n R$ is direct, we conclude $\delta_{n1} - cr_{n-1} + r_n - x_n = q_n \in w_n^o = \bigcap_{j=g(n-1)+1}^{g(n)} I_j \leq P_0 + c^n R$. Recursively we deduce $r_n = -c^{n-1} + \sum_{i=1}^{n} c^{n-i}(x_n + q_n)$. As $r_n = 0$ holds for large n, we obtain $c^{n-1} = \sum_{i=1}^{n} c^{n-i}(x_n + q_n) \in I_0 + \sum_{i=1}^{n} c^{n-i}(P_0 + c^i R) \leq P_0 + c^n R$. Since c is regular modulo P_0, we conclude $1 \in P_0 + cR \leq m$. This is the desired contradiction.

□

We are now able to establish the pivotal special case of the structure theorem.

Theorem 5.15. *Let* R *be a commutative noetherian ring. Consider a collection of discrete hollow* R*–modules, which are pairwise relatively projective, and have the same attached maximal ideal. Then their direct sum is discrete, if and only if it is finite, or all the summands are local and artinian.*

More specifically, and with the notation of Proposition 5.10, *a direct sum of discrete hollow* R*–modules with the same attached maximal ideal is discrete precisely in the following two cases: Either it involves finitely many copies of one local module* R/I *and of finitely many non–local modules* $(R/I_k)_{P_k}$, *subject to the further restrictions that* $I \leq \cap I_k$, *and that* R/I_k *is complete if* $(R/I_k)_{P_k}$ *appears with multiplicity greater than one. Or it involves infinitely many copies of one artinian local module* R/I.

PROOF. Everything follows immediately from Corollary 4.50, Theorem 4.53, Theorem 5.2, and the results of this section. We note specifically that Lemma 5.11 implies $I \leq \cap I_k$, if a local module R/I and non–local ones, $(R/I_k)_{P_k}$, do occur. Moreover, in this situation, $I \leq I_k \leq P_k$ and $\dim P_k = 1$ imply that R/I is not artinian; and therefore the multiplicity of R/I is finite. □

The general case, with more than one attached maximal ideal, reduces to this special one. For a direct sum M of hollow discrete modules, denote by M(m) the subsum of those with attached maximal ideal m. Then we have

<u>Corollary 5.16</u>. M *is discrete if and only if* M(m) *is discrete for all* m.

PROOF. Write $M = M_1 \oplus M_2$, where M_1 is the sum of certain M(m), and M_2 is the sum of the remaining ones. We claim that every homomorphism f between subfactors of M_1 and M_2 is zero.

Indeed, let $f(\bar{x}) = \bar{y}$. Then $x \in \bigoplus_{i \in F} H_i$ and $y \in \bigoplus_{j \in G} H_i$, for finite subsums of M_1 and M_2. Let $A = \bigcap_{i \in F} H_i^o$ and $B = \bigcap_{j \in G} H_j^o$. Then $xA = 0$ and $yB = 0$, and therefore $\bar{y}(A+B) = 0$. But, since the attached maximal ideals of the H_i and H_j are distinct, $A+B = R$. Thus $\bar{y} = 0$, and consequently $f = 0$.

With the claim at hand, the corollary follows immediately from Corollary 4.49.

(Instead of referring to Corollary 4.49, one can deduce from the claim that each submodule X of M decomposes as $X = \oplus (X \cap M(m))$. With this information, (D_1) and (D_2) follow directly for M, from their validity for the M(m).) □

<u>Remark</u>. The decomposition $M = \oplus M(m)$ of Corollary 5 is an instance of a generalized primary decomposition (Zöschinger [82b]). This decomposition is most naturally described in the language of torsion theories; cf. Stenström [75], Chapter VI, whose terminology we shall use.

Let R be any commutative ring. Call an ideal m–isolated if it is contained in at most one maximal ideal, m. The m–isolated ideals form a Gabriel topology, and therefore define a hereditary torsion class, T(m). For any R–module X, the sum of its T(m)–torsion submodules X(m) is direct. If $X = \oplus X(m)$, then $X(m) = X_m$, the localization of X at m.

The collection T of modules of the form $X = \oplus X(m)$ forms a hereditary pretorison class; the corresponding linear topology is generated by all isolated ideals (as a subbase). If R is noetherian, T is a hereditary torsion class.

To see that this decomposition $X = \oplus X(m)$ generalizes primary decomposition, note that Theorem 2.6 of Brandal [79] says that every torsion R–module belongs to T if and only if R is h–local. Every commutative ring of Krull dimension one, and in particular the ring of integers, is obviously h–local.

We conclude the section with a number of examples of non–local hollow discrete modules.

Example 5.17. Dedekind domains R.

To obtain a non–local hollow discrete module, $(R/I)_P$, one needs $P = I = 0$, and R local. Thus R must be a valuation ring, and then the quotient field is the only such module. It is quasi–projective if and only if the valuation ring is complete.

Example 5.18. A local ring R with many non–local hollow discrete modules.

Take R to be the localization, at the maximal ideal generated by x and y, of either the polynomial ring $\mathbb{R}[x,y]$ or the power series ring $\mathbb{R}[[x,y]]$, over the real numbers.

Each one–dimensional prime ideal of R is generated by a prime polynomial p(x,y) without constant term. As R_P is a discrete valuation ring, its ideals are the powers P_P^s, and the candidates for our modules are the $(R/P^s)_P$. In the second case, they will be quasi–projective. Such a candidate is indeed non–local hollow and discrete if and only if the integral closure of R/P is local; and this means that the curve $p(x,y) = 0$ has only one branch at (0,0), that is is either regular or has a hypercusp. For example, $x^3 - y^2$ leads to such a module, while $x^3 + x^2 - y^2$ does not.

Example 5.19. A ring R with infinitely many maximal ideals, each attached to non–local hollow discrete modules.

Take R to be the localization of either $\mathbb{R}[x,y]$ or $\mathbb{R}[x][[y]]$, at the complement of $\bigcup_{n\in N} \mathfrak{m}_n$, where N is any countable subset of $\mathbb{R}$, and $\mathfrak{m}_n$ is the maximal ideal generated by $x+n$ and y. As $\mathbb{R}$ is uncountable, the maximal ideals of R are precisely the localizations of the $\mathfrak{m}_n$ (cf. Jategaonkar [86], (7.1.5) and (7.2.6) or Müller [80], Proposition 21). The ideal P_{nk} of R generated by $x+n+y^k$, is prime and $\mathfrak{m}_n$–isolated. The factoring R/P_{nk} is a valuation ring, and complete in the second case. Therefore the modules $(R/P^s_{nk})_{P_{nk}}$, $(s \in \mathbb{N})$ are non–local hollow discrete with attached maximal ideal $\mathfrak{m}_n$.

COMMENTS

Theorem 5.2 appears in Mohamed, Müller and Singh [85].

The basic properties of the endomorphism ring of a discrete module are given in Mohamed and Singh [77]; the quasi–projective case is due to (Sandomierski and) Mares [63]. Lemma 5.8 and Corollary 5.9 are due to the authors.

Everything in the last section comes from Mohamed and Müller [88a].

APPENDIX

In this appendix, we discuss a number of topics which are related to the material of the book. Proofs will be given only where they are not readily available in the literature.

Several variants of supplementation and their interactions are considered. Cases where explicit structural information is available are compiled. Results on the splitting of supplements, and on modules with (C_1) are summarized. The historical origin of the concept of continuity, in von Neumann algebras and continuous geometries, is described. A brief overview is given of background and recent work on the weaker concept of $\aleph_0$–continuity. In the last section we list a number of open problems.

1. VARIANTS OF SUPPLEMENTATION

Recall that a module M is supplemented if, for every decomposition $M = A + B$, there exists a supplement of A contained in B; and that M has (D_1) if it is supplemented and every supplement is a summand.

Definitions A.1. A module M is *weakly supplemented* if every submodule has a supplement, ⊕*–supplemented* if every submodule has a supplement which is a summand, H*–supplemented* if for every submodule A there is a summand A' such that $A + X = M$ holds if and only if $A' + X = M$.

Proposition A.2. *The following implications hold:*

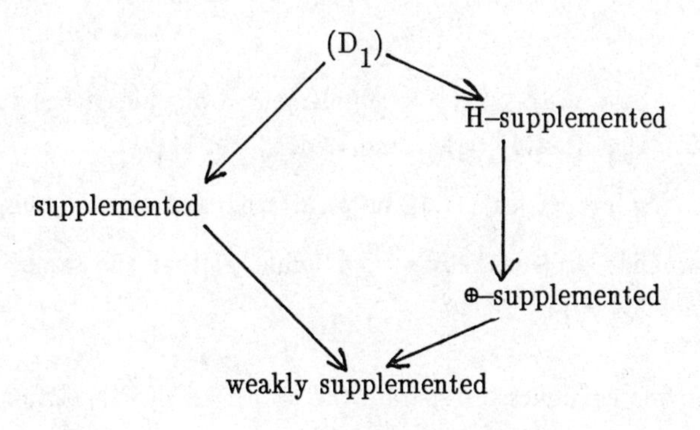

PROOF. (D_1) implies H–supplemented: Consider a submodule A of the module M. By (D_1) we have $M = M_1 \oplus M_2$, where $M_1 \leq A$ and $M_2 \cap A$ is small. We infer $A = M_1 \oplus M_2 \cap A$. If $M = A + X$, then $M = M_1 + M_2 \cap A + X$ hence $M = M1 + X$.

$M = A' \oplus B$, such that $M = A + X$ if and only if $M = A' + X$. In particular, $M = A + B$. If $M = A + B'$ for $B' \leq B$, then $A' + B' = M = A' \oplus B$, hence $B' = B$. We conclude that the summand B is a supplement of A.

The remaining implications are trivial. □

Next we investigate the relationship between H–supplementation and the lifting property.

<u>Proposition A.3</u>. *If* $\mathrm{Rad}M << M$, *then* M *is H–supplemented if and only if* $\overline{M} = M/\mathrm{Rad}M$ *is semisimple, and each summand (=submodule) of* $\overline{M}$ *lifts to a summand of* M.

PROOF. Given H–supplementation, consider $\overline{A} \leq \overline{M}$. The full inverse image A has a supplement B. Then $A \cap B$ is small in B hence in M. Therefore $A \cap B \leq \mathrm{Rad}M$, and consequently $\overline{A} \oplus \overline{B} = \overline{M}$. We conclude that $\overline{M}$ is semisimple.

Returning to $\overline{A} \leq \overline{M}$, we have a summand $A' \leq M$ such that $M = A + X$ if and only if $M = A' + X$. Consequently $\overline{M} = \overline{A} + \overline{X}$ if and only if $\overline{M} = \overline{A}' + \overline{X}$. As $\overline{M}$ is semisimple, this implies $\overline{A} = \overline{A}'$. Thus $\overline{A}$ lifts to the summand A'.

In the converse direction, if $N \leq M$ is given, then $M = M_1 \oplus M_2$ with $\overline{M}_1 = \overline{N}$. As RadM is small, we have $M = M_1 + X$ if and only if $M = N + X$. Thus M is H–supplemented. □

<u>Remarks</u>. (1) It is easy to see that a $\oplus$–supplemented module M has a decomposition $M = M_1 \oplus M_2$ such that $\mathrm{Rad}M_1 = M_1$ and $\mathrm{Rad}M_2 << M_2$.

(2) While the properties (D_1), supplemented and weakly supplemented are inherited by summands, it is unknown (and unlikely) that the same is true for the other two properties.

We now provide examples to separate the properties of Proposition A.2.

<u>Lemma A.4</u>. *Let* R *be a commutative local ring, with maximal ideal* J, *and let* M *be a finitely generated* R*–module. Then*

(1) M *is supplemented;*

(2) M *is* ⊕*–supplemented if* $M \cong \bigoplus_{i=1}^{n} R/I_i$, *for ideals* I_i;

(3) M *is* H*–supplemented if and only if, in addition,* $I_1 \leq \ldots \leq I_n$;

(4) M *has* (D_1) *if and only if, in addition,* $JI_n \leq I_1 \leq \ldots \leq I_n$;

(5) M *is (quasi–)discrete if and only if, in addition,* $I_1 = \ldots = I_n$.

If M *is 2–generated, then* (2) *reads "if and only if" also.*

PROOF. Everything but the last statement is proved in Zöschinger [82a]. Let M be 2–generated and ⊕–supplemented, and pick $a \in M - \mathrm{Rad}M$. Then aR has a supplement which is a summand, i.e. $M = M_1 \oplus M_2$ with $aR \cap M_2$ small. Thus $\overline{M} = \overline{aR} \oplus \overline{M}_2$ with $0 \neq \overline{aR} \cong \overline{M}_1$. Therefore the $\overline{M}_i$ are both 1–dimensional over the field R/J, and consequently $M_i \cong R/I_i$ with $I_i \leq J \leq R$. □

<u>Lemma A.5</u>. *Let* R *be an incomplete rank one discrete valuation ring, with quotient field* K. *Then the module* $M = K^2$ *is* ⊕*–supplemented but not supplemented.*

PROOF. By Zöschinger ([74b], 2.2) every submodule of M has a supplement, but M is not supplemented. Let B be a supplement of A in M. As M is divisible, $M = Mr = Ar + Br \leq A + Br$, for every $0 \neq r \in R$. Thus B = Br, and B is divisible, hence injective, hence a summand. □

<u>Corollary A.6</u>. *No other implications hold in the diagram of* Proposition 2, *except possibly* "H–supplemented ⇒ supplemented".

PROOF. Use Lemmas 4 and 5. □

We now provide a complete list of all the relevant modules over Dedekind domains. All these results are due to Zöschinger [74a], [74b], [82a]; cf. also Hausen [82], Mohamed and Abdul–Karim [84], and Singh [84].

<u>Proposition A.7</u>. *Let* R *be a local Dedekind domain, with maximal ideal* P, *quotient field* K, *and* Q = K/R. *Let* a, b, c *and* n *be natural numbers, and let* $B(n_1,\ldots,n_s)$ *denote the direct sum of arbitrarily many copies of* $R/P^{n_1},\ldots,R/P^{n_s}$. *Then the following table describes all modules* M *with the listed properties:*

Property	Structure	
weakly supplemented, $\oplus$–supplemented	$M \cong R^a \oplus K^b \oplus Q^c \oplus B(1,\dots,n)$	
supplemented, H–supplemented	M as above	($b \leq 1$ if R is incomplete)
(D_1)	$M \cong R^a \oplus K^b \oplus Q^c$ or $B(n,n+1)$	
quasi–discrete	$M \cong R^a \oplus K^b$ or $R^a \oplus Q$ or $B(n)$	
discrete	$M \cong R^a \oplus K^b$ or $B(n)$	

PROOF, with indication of original terminology: Zöschinger [74a], 2.4 and [82a], 2.10 "komplementiert"; [74b], 2.2 and [82a], 2.6 "supplementiert"; [74b], 4.3 "stark komplementiert"; [74b], 5.2 "stark supplementiert"; Theorem 5.2. □

<u>Proposition A.8</u>. *If* R *is a non–local Dedekind domain, then all these modules are torsion. Moreover, a torsion module has any one of these properties if and only if every* P*–primary component (viewed as module over the localization* R_P*) has the structure described in* Proposition A.7. (Note that "torsion" forces $a = b = 0$, and that c and n will vary with P.)
PROOF. Zöschinger [74a], 3.1. □

<u>Remark</u>. Rudlof [89], a student of Zöschinger, has recently fully determined the structure of all weakly supplemented modules over commutative noetherian rings.

2. SUPPLEMENTS ARE SUMMANDS

The property of the title has been widely studied for projective modules. We summarize the main results:

<u>Proposition A.9</u>. *The following are equivalent, for any ring:*

(1) *In every finitely generated projective right module, every supplement submodule is a summand;*

(2) *every projective right module with finitely generated radical factor module, is itself finitely generated;*

(3) *in every projective right module, every finitely generated submodule is contained in a maximal submodule;*

(4) *the analogues of the above properties, for left modules.*

All these properties are true for commutative rings, as well as for rings such that every prime factor ring is right (or left) Goldie *(in particular for right or left noetherian rings, and for rings with a polynomial identity).*

If R *is a commutative domain or a right noetherian ring, then* (1) *is valid for every (not necessarily finitely generated) projective module.*

PROOF. Cf. Zöschinger [81] and the literature cited there, as well as Mohammed and Sandomierski [P], where several other equivalent conditions are listed. □

An example of a ring where these equivalent conditions fail, is provided in Gerasimov and Sakhaev [84]; cf. also Mohammed and Sandomierski [P] for a discussion of "minimal counterexamples".

3. EXTENDING MODULES

We summarize briefly some of the results in Harada [82b], Okado [84] and Kamal and Müller [88a,b,c], concerning the structure of modules with (C_1) (which are also called *extending modules*), over certain commutative rings. Many questions remain open here.

Proposition A.10. *Let* R *be a commutative domain. A module* M *has* (C_1) *if and only if it is a torsion module with* (C_1), *or the direct sum of an injective module, and a torsionfree reduced module with* (C_1).

If M *is torsionfree reduced, then it has* (C_1) *if and only if it is a finite direct sum of uniform submodules, each pair of which has* (C_1). □

Proposition A.11. *Let* R *be a commutative noetherian ring. If a module* M *has* (C_1), *then it is the direct sum of uniform submodules.*

If $M = \oplus M_i$ *where all* M_i *have local endomorphism rings, then* M *has* (C_1) *if and only if it satisfies* lsTn, *and every pair* $M_i \oplus M_j$ *has* (C_1). □

(Fairly complicated) criteria as to when the direct sum of two uniform modules has (C_1), are contained in Kamal and Müller [88b,c]. Complete results are known only over Dedekind domains:

Proposition A.12. *Let* R *be a Dedekind domain, with maximal ideals* P. *A torsion module has* (C_1) *if and only if each* P*–primary component is a direct sum of copies of either* C_P^{∞}, *or of* C_P^n *and* C_P^{n+1} *for some natural number* $n = n(P)$. *A torisonfree reduced module has* (C_1) *if and only if it is a finite direct sum* $\oplus NI_i$, *where* N *is a torsionfree reduced uniform module* (i.e., a proper submodule of the quotient field of R), *and the* I_i *are fractional ideals.* □

4. THE HISTORICAL ORIGIN OF THE CONCEPT OF CONTINUITY

Before we start our discussion, we remind the reader of a few concepts from lattice theory.

A complete lattice is *upper* [*lower*] *continuous* if $x \wedge (\vee x_i) = \vee(x \wedge x_i)$ $[x \vee (\wedge x_i) = \wedge(x \vee x_i)]$ holds for any chain or up [down] directed family $\{x_i\}$.

A lattice is *modular* if $x \leq z$ implies $x \vee (y \wedge z) = (x \vee y) \wedge z$.

A lattice with 0 and 1 is *complemented* if for every element x there is a complement, ie., an element y with $x \wedge y = 0$ and $x \vee y = 1$. Two elements are *perspective* if they have a common complement.

An *orthocomplemented lattice* is a lattice with an additional operation, ′, called *orthocomplementation*, which is involutory and anti–monotone, and such that x′ is a complement of x.

An orthocomplemented lattice is *orthomodular* if $x \leq z$ implies $x \vee (x' \wedge z) = z$.

In quantum mechanics (von Neumann [32]), the physical observables are represented by selfadjoint operators, $a = a^*$, on a complex separable Hilbert space. Two observables are simultanuously measurable if and only if their operators commute.

The analysis of such operators let Murray and von Nemann [36] to investigate what are now called *von Neumann algebras*, i.e., *–subalgebras of the algebra of all bounded operators, which coincide with their second commutator.

Von Neumann algebras have two fundamental properties: Every left or right annihilator is generated by a *projection* (ie. a self adjoint idempotent $e = e^* = e^2$). The set PA of all projections forms a complete orthomodular lattice. (Here, PA is partially ordered via $e \leq f$ iff $e = ef$, and the orthocomplementation is defined as $e' = 1-e$.) Elements $e, f \in PA$ are called *equivalent*, $e \sim f$, if there is $a \in A$ with $e = aa^*$ and $f = a^*a$. An element $e \in PA$ is *finite* if $e \sim f \leq e$ implies $e = f$.

A *factor* is the special case of a von Neumann algebra whose center consists of the complex numbers only. Factors are precisely those von Neuman algebras which occur in tensor decompositions $A_1 \otimes A_2$ of the algebra of all bounded operators. For a factor A, the collection of equivalence classes in PA is totally ordered.

Moreover there exists a *dimension function* (unique up to a scalar multiple) $d : PA \longrightarrow \mathbb{R}^+ \cup \{\infty\}$ satisfying the following properties: $de = 0$ iff $e = 0$, $de = df$ iff $e \sim f$, $de < \infty$ iff e is finite, $d(e+f) = de + df$ if $ef = 0$.

The range of this dimension function is limited to the following possibilities : $\{0,1,2,...,n\}$, $\{0,1,2,...,\infty\}$, $[0,1]$, $[0,\infty]$ and $\{0,\infty\}$. Accordingly, Murray and von Neumann classified von Neumann algebras into the five types I_n, I_∞, II_f, II_∞ and III. All types occur. The orthomodular lattice PA is modular precisely for the two types I_n and II_f.

Generalizing these last two cases, von Neumann ([36a], [60]) defined a "geometry" to be a complete complemented modular lattice L which is upper and lower continuous (and indecomposable). Kaplansky [55] established that a complete orthocomplemented modular lattice is automatically upper and lower continuous (cf. also Amemiya and Halperin [50]).

With considerable effort (mainly devoted to showing that perspectivity is transitive) von Neumann found, in the indecomposable case, an (essentially unique) dimension function $d : L \longrightarrow R^+$ satisfying $dx = 0$ iff $x = 0$, $d(x \vee y) = dx + dy$ if $x \wedge y = 0$. Its range is limited to the two cases $\{0,1,2,...,n\}$ and $[0,1]$. In the first instance, L is just an (n–1)–dimensional *projective geometry* (cf. Birkhoff [35]). In the second one, L is called a *continuous geometry.* (The decomposable case is technically more complicated and requires a family of dimension functions, cf. Iwamura [44]).

Next von Neumann [36c] defined regular rings (a ring R is *regular* if for every $x \in R$ there is $y \in R$ with $xyx = x$). He showed that, if a geometry L has at least four perspective independent elements, then there is a (unique) regular ring R whose lattice of principal right ideals is isomorphic to L. In the projective case, R is artinian, in fact an $n \times n$–matrix ring over a division ring. In the continuous case, R is not artinian, and is called a *continuous* regular ring.

A *right continuous* regular ring is one whose lattice of principal right ideals is upper continuous. As the lattices of principal left and right ideals are dual to each other, under annihilators, a regular ring is continuous if and only if it is left and right continuous.

We present now the proof, due to Utumi [61], that a regular ring is right continuous if and only if it satisfies the condition (C_1). Note that the condition (C_2) is always satisfied for a regular ring.

Lemma A.13. *The principal right ideals of a regular ring form a sublattice of the lattice of all right ideals.*

PROOF. Recall that a ring R is regular if and only if for every $a \in R$ there exists $x \in R$ with $axa = a$. It follows immediately that ax is idempotent, and $aR = axR$. Thus a right ideal is principal if and only if it is a summand. (We shall use this fact repeatedly in the sequel.)

Consider two right ideal A, B of R which are summands. Let $A \oplus A^* = R$, and let π denote the projection onto A^*. We obtain $A + B = A \oplus \pi B$. Moreover πB is principal hence a summand, of R and therefore of A^*. We conclude $A + B = A \oplus \pi B \subset^{\oplus} R$.

Since B is a summand hence projective, the epimorphism $B \longrightarrow\!\!\!\!\to B/A \cap B \cong A + B/A \cong \pi B$ splits. We deduce that $A \cap B$ is also a summand. □

Proposition A.14. *A regular ring is right continuous if and only if it satisfies* (C_1) *for right ideals.*

PROOF. Assume that R is right continuous, and consider an arbitrary right ideal A. By completeness, the join $\bigvee_{a \in A} aR$ in the lattice of principal right ideals exists, but is possibly larger than the join $\sum_{a \in A} aR = A$ in the lattice of all right ideals. We establish (C_1) by showing $A \leq^e \bigvee_{a \in A} aR$.

To this end consider $B \leq \bigvee_{a \in A} aR$ with $B \cap A = 0$. For a finite subset F of A, $\bigvee_{a \in F} aR = \sum_{a \in F} aR$ holds by Lemma A.13. Therefore $B \cap \bigvee_{a \in F} aR = B \cap \sum_{a \in F} aR \leq B \cap A = 0$. Then, by upper continuity, and since the $\bigvee_{a \in F} aR$ form an updirected family, $B \cap \bigvee_{a \in A} aR = B \cap \bigvee_F (\bigvee_{a \in F} aR) = \bigvee_F (B \cap \bigvee_{a \in F} aR) = 0$. We conclude $B = 0$, as required.

Conversely suppose that R satisfies (C_1). Let $\{A_i\}$ be a chain of summands. By (C_1), the right ideal ΣA_i is essential in a summand A. We demonstrate completeness by showing that A is the least upper bound for $\{A_i\}$, in the lattice of summands.

Let $B \subset^{\oplus} R$ be any other upper bound for $\{A_i\}$. Then $A \cap B \subset^{\oplus} R$ by Lemma A.13, and hence $A = A \cap B \oplus C$ for some right ideal C. We obtain $\Sigma A_i \cap C \leq A \cap B \cap C = 0$, and therefore $C = 0$ since ΣA_i is essential in A. Consequently $A = A \cap B$, and $A \leq B$, as required.

It remains to show upper continuity. Let D be an arbitrary summand. Then $D \cap A_i$ is a summand, and $\vee(D \cap A_i)$ exists by completeness. We obtain $\Sigma (D \cap A_i) \leq \vee(D \cap A_i) \leq D \cap \vee A_i$. Since $\Sigma A_i \leq^e \vee A_i$ holds, as seen before, we deduce $\vee(D \cap A_i) \leq^e D \cap \vee A_i$. But both terms are summands, and therefore equal. □

The study of von Neumann algebras has led to a tremendous amount of literature; for instance, Dixmier [81] lists 968 papers till 1980. An excellent survey of the related work on orthomodular lattices, till 1966, can be found in Holland [70]; cf. also Kalmbach [83].

5. $\aleph_0$–CONTINUOUS RINGS AND MODULES

$\aleph_0$–continuous regular rings have arisen in two different, though somewhat related contexts:

First, Halperin [38] showed that most of von Neumann's results on continuous geometries could be obtained for $\aleph_0$–continuous geometries (ie. $\aleph_0$–complete $\aleph_0$–upper and lower continuous complemented modular lattices). In particular, most $\aleph_0$–continuous geometries can be coordinatized by "$\aleph_0$–continuous" regular rings. In the spirit of Utumi (cf. Section 4 of this appendix), a regular ring is $\aleph_0$–continuous if and only if it is left and right $\aleph_0$–continuous, and it is right $\aleph_0$–continuous if and only if it satisfies ($\aleph_0$–C_1): every $\aleph_0$–generated right ideal is essential in a summand.

For the discussion of the second context, we require some background. It was felt early on that it was desirable to develop the theory of von Neumann algebras (= W^*–algebras = weakly closed $*$–subalgebras of the algebra of all bounded operators on a Hilbert space), as far as possible, in a purely algebraic–topological framework, without reference to operator theory. Kaplansky [51] introduced AW^*–algebras as an abstract generalization of W^*–algebras.

We describe the relevant concepts in the setting of C^*–algebras (ie. complex Banach algebras with involution subject to $\|xx^*\| = \|x\|^2$, equivalently norm–closed

*–subalgebras of the algebra of all bounded operators on a Hilbert space, I. Gelfand and M. Naimark, cf. Goodearl [82a]), though some of the arguments can be carried out in the purely algebraic setting of *–rings or even ordinary rings. An AW^*–algebra [Rickart C^*–algebra] is a C^*–algebra such that the right (or left) annihilator of every subset [element] is generated by a projection.

The set PA of projections of such an algebra A is an [$\aleph_0$–]complete orthocomplemented lattice. If A is finite (ie. $xx^* = 1$ implies $x^*x = 1$), then PA is modular, and is in fact an [$\aleph_0$–]continuous geometry. This geometry can always be coordinatized, by an [$\aleph_0$–]continuous regular ring R. Interestingly enough it has turned out recently that this ring R is actually the maximal left and right [classical] quotient ring of A (cf. Handelman and Lawrence [78], Ara and Menal [84].)

$K_0(R)$, for $\aleph_0$–continuous regular rings, is investigated in great detail in Goodearl [79]. Crucial facts are that such rings are unit regular, and that their finitely generated projective modules have cancellation and interpolation properties. Some structural information on R is deduced. Many of these results extend to directly finite right $\aleph_0$–continuous regular rings (Goodearl [82b]), and even to arbitrary right $\aleph_0$–continuous regular rings (Ara [87]). In Goodearl, Handelman and Lawrence [80] they use the embedding of a finite Rickart C^*–algebra A into an $\aleph_0$–continuous regular ring R very effectively to transfer many of these results from R to A.

A few recent papers attempt to extend some of this work from right $\aleph_0$–continuous regular rings and their finitely generated projective modules, to (more or less) arbitrary $\aleph_0$–(quasi–)continuous modules over (more or less) arbitrary rings. The definitions (which rest on appropriate $\aleph_0$–relativizations of the conditions (C_i)) and elementary properties are given by Oshiro [83b]. Kutami [83], working over arbitrary regular rings, proves an analogue of (2.32) for $\aleph_0$–essentially generated (ie. essential over an $\aleph_0$–generated subsubmodule) submodules of a directly finite $\aleph_0$–quasi–continuous projective module.

Mucke [88] has quite a number of more general results. Over arbitrary rings, he proves analogues of (3.18) for $\aleph_0$–(quasi–)continuous $\aleph_0$–essentially generated modules with (split) embeddings into each other. He also establishes the cancellation property for directly finite, and the finite interpolation property for arbitrary $\aleph_0$–continuous $\aleph_0$–essentially generated modules with the finite exchange property.

More penetrating results are obtained over two types of rings, viz. right coherent rings and right ZP–rings (ie. rings where the right annihilator of every element is generated by an idempotent), and for $\aleph_0$–(quasi–)continuous $\aleph_0$–essentially generated submodules of free modules. For instance, in the $\aleph_0$–quasi–continuous case, such modules satisfy the internal cancellation property, the $\aleph_0$–interpolation property, and analogues of (2.32), (3.14) and (2.14). In the $\aleph_0$–continuous case, the finite exchange property is valid, and therefore the cancellation property if the module is directly finite.

6. OPEN PROBLEMS

1. Characterize the quasi–continuous modules with the finite exchange property, cf. (3.24) and (2.25). Using (2.37) and (3.20), one can confine oneself to square free modules M. Such a module might have the finite exchange property iff end M has a "sheaf representation" with local stalks.)

2. Does the finite exchange property imply the full exchange property, for quasi–continuous modules ?

3. Characterize the cancellation property for quasi–continuous modules (cf. (1.23), (2.33) and (3.25)).

4. Determine the rings over which every continuous module is quasi– injective (see Rizvi [88] for partial results).

5. Find analogues of the theorems of Chapters 2 and 3 for $\aleph_0$–(quasi–) continuous modules (for numerous partial results, see Mucke [88], cf. also Section 5 of this appendix).

6. Investigate the structure of ($\aleph_0$–,quasi–)continuous rings, in generalization of ($\aleph_0$–)continuous regular rings; cf. Section 5 of the Appendix.

7. Study modules and rings with the condition (C_1) only, generalizing the results in Kamal and Müller [88a,b,c] and Chatters et al. [77, 80].

8. In particular, find necessary and sufficient conditions, for a module with (C_1) to be the direct sum of indecomposables (for a sufficient condition, cf. (2.17) – (2.19)).

9. Weaken (C_1) to the analogue of ⊕–supplemented: every submodule has a complement which is a summand.

10. Develop a fully satisfactory structure theory for injective modules over an arbitrary ring, along the ideas in Goodearl and Boyle [76], and its consequences for the structure of quasi–continuous modules.

11. The endomorphism ring of a local R–module is a local ring, provided R is commutative or right noetherian. Investigate this implication in general. Find an example where it fails. Find an infinite collection of local modules with non–local endomorphism rings which are relatively projective in pairs (cf. (4.53)).

12. Can the conditions of (4.48) be weakened ? Are (b) and (c) independent, in the presence of (a) ? Is it sufficient to assume only (a) and lsTn ? (All of this is of interest only if the end M_i are not local; cf. (4.49) and (2.25).)

13. In the situation of (4.53), every local summand is a summand, by (4.13). Is this condition also equivalent to the conditions (1) – ((5) of the list? (In showing this, one may assume that the M_i are pairwise non–isomorphic.)

14. Determine the structure of (quasi–)discrete modules over arbitary rings. (In analogy with (5.15), (5.16) one might expect that any (quasi–) discrete module has a "primary decomposition" such that each component is either a finite direct sum of hollow modules, or a direct sum of local modules. These two cases are covered by (4.50) and (4.53).)

15. Does the exact dual of (3.2) hold; ie. is (5.8) valid for supplemented modules, except for the phrase "and is a summand of M"?

16. Determine when a quasi–discrete module has the cancellation property (cf. (1.23) and (4.20)).

17. Characterize the rings over which every discrete module is quasi–projective. (This reduces immediately to studying hollow modules.)

18. Is every H–supplemented module supplemented (cf. (A.2))?

19. Are the properties H–supplemented and ⊕–supplemented inherited by summands?

20. Does the converse of (A.4) (2) hold for $n > 2$?

21. Investigate the structure of (weakly) supplemented modules over arbitrary rings (cf. the forthcoming results of P. Rudlof over commutative noetherian rings, and Inoue [83]).

22. Are existing supplements in arbitrary projective modules summands, over any commutative ring (cf. (A.9)) ?

23. Characterize the rings of (A.9) internally.

24. For continuous modules, one knows $\Delta = J(S)$, cf. (3.5) and (3.15); for discrete modules, $\nabla = J(S)$, cf. (5.4). Investigate the relationship between Δ, ∇ and $J(S)$, for quasi–continuous and quasi–discrete modules.

BIBLIOGRAPHY

Abdul–Karim, F. H. [82]: Contributions to the theory of dual–continuous modules, M. Sc. thesis, Univ. Kuwait, (1982).

Ahsan, J. [73]: Rings all whose cyclic modules are quasi–injective, Proc. London Math. Soc. **27**, (1973), 425–439.

_______ [81]: Submodules of quasi–projective modules, Comm. Algebra **9**, (1981), 975–988.

Amemiyah, I. and Halperin, I. [50]: Complemented modular lattices, Canad. J. Math. **11**, (1959), 481–520.

Anderson, D. D. [76]: The existence of dual modules, Proc. Amer. Math. Soc. **55**, (1976), 258–260.

Anderson, F. W. [69]: Endomorphism rings of projective modules, Math. Z. **111**, (1969), 322–332.

Anderson, F. W. and Fuller, K. R. [72]: Modules with decompositions that complement direct summands, J. Algebra **2**, (1972), 241–253.

_______ [73]: Rings and Categories of Modules, Springer Verlag, 1973.

Ara, P. [87]: Aleph–nought–continuous regular rings, J. Algebra **109**, (1987), 115–126.

Ara, P. and Menal, P. [84]: On regular rings with involution, Arch. Math. **42,** (1984), 126–130.

Azumaya, G. [P]: M–projective and M–injective modules, unpublished (1974).

_______ [74]: Characterization of semiperfect and perfect modules, Math. Z. **140**, (1974), 95–103.

Azumaya, G., Mbuntum, F. and Varadarajan, K. [75]: On M–projective and M–injective modules, Pacific J. Math. **95**, (1975), 9–16.

Baer, R. [40]: Abelian groups that are direct summands of every containing abelian group, Proc. Amer. Math. Soc. **46**, (1940), 800–806.

Bass, H. [60]: Finitistic dimension and homological generalization of semiprimary rings, Trans. Amer. Math. Soc. **95**, (1960), 466–488.

_______ [62]: Injective dimension in noetherian rings, Trans. Amer. Math. Soc. **102**, (1962), 18–29.

Beck, I. [78]: On modules whose endomorphism ring is local, Israel J. Math. **29**, (1978), 393–407.

_______ [78]: An independence structure on indecomposable modules, Ark. Mat. **16**, (1978), 171–178.

Berberian, S. K. [57]: The regular ring of a finite AW^*–algebra, Ann. Math. **65**, (1957), 224–240.

Birjukov, P. A. [78]: On a certain class of Abelian torsion groups, Notices Amer. Math. Soc. **25**, (1978), A–418.

Birkenmeier, G. F. [76]: On the cancellation of quasi–injective modules, Comm. Algebra **4**, (1976), 101–109.

_______ [76b]: Self–injective rings and the minimal direct summand containing the nilpotents, Comm. Algebra **4**, (1976), 705–721.

_______ [78]: Modules which are subisomorphic to injective modules, J. Pure Applied Algebra **13**, (1978), 169–177.

_______ [81]: Baer rings and quasi–continuous rings have a MDSN, Pacific J. Math. **97**, (1981), 283–292.

_______ [83]: Modules which are epi–equivalent to projective modules, Acta Mathematica **24**, (1983), 9–16.

_______ [P]: Quasi–projective modules and the finite exchange property.

Birkhoff, G. [35]: Combinatorial relations in projective geometries, Ann. Math. **36**, (1935), 743–748.

Björk, J.–E. [70]: Rings satisfying a minimal condition on principal ideals, J. Reine Angew. Math. **245**, (1970), 63–73.

_______ [72]: Radical properties of perfect modules, J. Reine Angew. Math. **253**, (1972), 78–86.

Bouhy, T. [75]: On quasi–injective modules, M.Sc. thesis, Ain Shams Univ., Cairo, (1975).

Boyle, A. K. [74]: Hereditary QI–rings, Trans. Amer. Math. Soc. **192**, (1974), 115–120.

Boyle, A. K. and Goodearl, K. R. [75]: Rings over which certain modules are injective, Pacific J. Math. **58**, (1975), 43–53.

Brandal, W. [79]: Commutative Rings Whose Finitely Generated Modules Decompose, Lecture Notes in Mathematics **723**, Springer Verlag, 1979.

Brodskii, G. M. [83]: Dualities in modules and the AB^* condition, Russian Math. Surv. **38**, (1983), 185.

Bumby, R. T. [65]: Modules which are isomorphic to submodules of each other, Arch. Math. **16**, (1965), 184–185.

Burgess, W. D. and Raphael, R. [P]: On modules with the absolute direct sum property, unpublished (1985).

Byrd, K. A. [72]: Rings whose quasi–injective modules are injective, Proc. Amer. Math. Soc. **33**, (1972), 235–240.

_______ [79]: Right self–injective rings whose essential right ideals are two–sided, Pacific J. Math. **82**, (1979), 23–41.

Cartan, H. and Eilenburg, S. [56]: Homological Algebra, Princeton Univ. Press, 1956.

Chambless, L. [81]: Coprimary decomposition, N–dimension and divisibility: Application to artinian modules, Comm. Algebra **9**, (1981), 1131–1146.

Chase, U. S. [60]: Direct product of modules, Trans. Amer. Math. Soc. **97**, (1960), 457–473.

Chatters, A. W. and Hajaranavis, C. R. [77]: Rings in which every complement right ideal is a direct summand, Quart. J. Math. Oxford **28**, (1977), 61–80.

Chatters, A. W. and Khuri, S. M. [80]: Endomorphism rings of modules over non–singular CS rings, J. London Math. Soc. **21**, (1980), 434–444.

Crawley, P. and Dilworth, R. P. [73]: Algebraic Theory of Lattices, Prentice–Hall, 1973.

Crawley, P. and Jonsson, B. [64]: Refinements for infinite direct decompositions of algebraic systems, Pacific J. Math. **14**, (1964), 797–855.

Cunningham, R. S. and Rutter, E. A. Jr. [74]: Perfect modules, Math. Z. **140**, (1974), 105–110.

Diximer, J. [81]: Von Neumann Algebras, North Holland, 1981.

Eckmann, B. and Schopf, A. [53]: Über injective Moduln, Arch. Math. **4**, (1953), 75–78.

Faith, C. [73a]: When are proper cyclics injective? Pacific J. Math. **45**, (1973), 97–112.

_______ [73b]: Algebra: Rings, Modules and Catagories I, Springer Verlag, 1973.

_______ [76a]: Algebra II, Ring Theory, Springer Verlag, 1976.

_______ [76b]: On hereditary rings and Boyle's conjecture, Arch. Math. **27**, (1976), 113–119.

_______ [85]: The maximal regular ideal of self–injective and continuous rings splits off, Arch. Math. **44**, (1985), 511–521.

Faith, C. and Utumi, Y. [64]: Quasi–injective modules and their endomorphism rings, Arch. Math **15**, (1964), 166–174.

Faith, C. and Walker, E. A. [67]: Direct sum representation of injective modules, J. Algebra **5**, (1967), 203–221.

Faticoni, T. G. [83]: On quasi–projective covers, Trans. Amer. Math. Soc. **278**, (1983), 101–113.

Feigelstock, S. and Raphael, R. [85]: A problem of relative projectivity for abelian groups, Canad. Math. Bull. **29**, (1986), 114–122.

_______ [86]: Some aspects of relative projectivity, Comm. Algebra **13**, (1986), 1187–1212.

Fieldhouse, D. [84]: Epis and monos which must be isos, Internat. J. Math. & Math. Sci. **7**, (1984), 507–512.

Fleury, P. [74a]: A note on dualizing Goldie dimension, Canad. Math. Bull, **17**, (1974), 511–517.

_______ [74b]: Hollow modules and local endomorphism rings, Pacific J. Math. **53**, (1974), 379–385.

_______ [77]: On local QF–rings, Aequat. Math. **16**, (1977), 173–179.

Fuchs, L. [69]: On quasi–injective modules, Annali Scoula Normal Sup. Pisa **23**, (1969), 541–546.

_______ [72]: The cancellation property for modules, in Lecture Notes Math. **246**, Springer Verlag, 1972, 193–212.

Fuchs, L., Kertesz, A. and Szele, T. [53]: Abelian groups in which every serving subgroup is a direct summand, Publ. Math. Debrecen **3**, (1953), 95–105.

Fuller, K. R. [69]: On direct representations of quasi–injectives and quasi–projectives, Arch. Math. **20**, (1969), 495–502.

Fuller, K. R. and Hill, D. A. [70]: On quasi–projective modules via relative projectivity, Arch. Math **21**, (1970), 369–373.

Generalov, A. I. [83]: The ω–cohigh purity in a category of modules, Math. Notes **33**, (1983), 402–408.

Gerasimov, V. A. [82]: Localizations in associative rings, Sib. Math. J. **23**, (1982), 788–804.

Gerasimov, V. N. and Sakhaev, I. I. [84]: A counter–example to two hypotheses on projective and flat modules, Sibirskii Mat. J. **25**, (1984), 31–35.

Goel, V. K., Jain, S. K. and Singh, S. [75]: Rings whose cyclic modules are injective or projective, Proc. Amer. Math. Soc. **53**, (1975), 16–18.

Goel, V. K. and Jain, S. K. [76]: Semiperfect rings with quasi–projective left ideals, Math. J. Okayama Univ. **19**, (1976), 39–43.

—— [78]: π–injective modules and rings whose cyclic modules are π–injective, Comm. Algebra **6**, (1978), 59–73.

Golan, J. S. [70]: Characterization of rings using quasi–projective modules, Israel J. Math. **8**, (1970), 34–38.

_______ [71a]: Characterization of rings using quasi–projective modules (II), Proc. Amer. Math. Soc. **2**, (1971), 237–343.

_______ [71b]: Quasi–perfect modules, Quart. J. Math. Oxford **22**, (1971), 173–182.

Goodearl, K. R. [76]: Direct sum properties of quasi–injective modules, Bull. Amer. Math. Soc. **82**, (1976), 108–110.

_______ [79]: Vov Neumann regular rings, Pitman, 1979.

_______ [82a]: Notes on Real and Complex C^*–Algebras, Shira Publ. Ltd., 1982.

_______ [82b]: Directly finite aleph–nought–continuous regular rings, Pacific J. Math. **100**, (1982), 105–122.

Goodearl, K. R. and Boyle, A. K. [76]: Dimension Theory for nonsingular injective modules, Memoirs Amer. Math. Soc. **177**, (1976).

Goodearl, K. R., Handelman, D. E. and Lawrence, J. W. [80]: Affine representations of Grothendieck groups and applications to Rickart C^*–algebras and $\aleph_0$–continuous regular rings, Memoirs Amer. Math. Soc. **234**, (1980).

Grzeszczuk, P. and Puczylowski, A. R. [84]: On Goldie and dual Goldie dimension, J. Pure Applied Algebra **31**, (1984), 47–54.

Gupta, A. K. and Varadarajan, K. [80]: Modules over endomorphism rings, Comm. Algebra **8**, (1980), 1291–1333.

Guralnik, R. M. [86]: Power cancellation of modules, Pacific J. Math. **124**, (1986), 131–144.

Haack, J. K. [82]: The duals of the Camillo–Zelmanowitz formulas for Goldie dimension, Canad. Math. Bull. **25**, (1982), 325–334.

Hafner, I. [74]: The regular ring and the maximal ring of quotients of a finite Baer *–ring, Michigan Math. J. **21**, (1974), 153–160.

Hall, P. [37]: Complemented groups, J. London Math. Soc. **12**, (1937), 201–204.

Halperin, I. [38]: On the transitivity of perspectivity in continuous geometries, Trans. Amer. Math. Soc. **44**, (1938), 537–562.

Handelman, D. [79]: Finite Rickart C^*–algebras and their properties, Studies in Analysis, Adv. Math. Suppl. Studies **4**, (1979), 171–196.

Handelman, D. E. and Lawrence, J. W. [78]: Lower K–theory, regular rings and operator algebras – a survey, in Lecture Notes in Math. **734**, Springer Verlag, 1978, 158–173.

Hanna, A. and Shamsuddin, A. [83]: On the structure of certain types of abelian groups, Arch. Math. **40**, (1983), 495–502.

Harada, M. [75]: On the exchange property of a direct sum of indecomposable modules, Osaka J. Math. **12**, (1975), 719–736.

_______ [77]: Small submodules in a projective module and semi–T–nilpotent sets, Osaka J. Math. **14**, (1977), 355–364.

_______ [78]: A note on hollow modules, Rev. Union Math. Argentina **28**, (1978), 186–194.

_______ [78b]: A note on hollow modules, Rev. Un. Mat. Argentina **28**, (1978), 186–194.

_______ [78b]: On small homomorphism, Osaka J. Math. **15**, (1978), 365–370.

_______ [78c]: On the small hulls of a commutative ring, Osaka J. Math. **15**, (1978), 679–682.

_______ [79]: Non–small modules and non–cosmall modules, in Ring Theory, Lecture Notes in Pure and Applied Math. **51**, Marcel Dekker, 1979.

_______ [80]: On lifting property on direct sums of hollow modules, Osaka J. Math. **17**, (1980), 783–791.

_______ [82a]: On modules with lifting properties, Osaka J. Math. **19**, (1982), 189–201.

_______ [82b]: On modules with extending property, Osaka J. Math. **19**, (1982), 203–215.

_______ [82c]: Uniserial rings and lifting properties, Osaka J. Math. **19**, (1982), 217–229.

_______ [83a]: Factor Categories with Applications to Direct Decomposition of Modules, Marcel Dekker, 1983.

_______ [83b]: On maxi–quasiprojective modules, J. Austral. Math. Soc. **35**, (1983), 357–368.

Harada, M. and Ishii, T. [72]: On endomorphism rings of noetherian quasi–injective modules. Osaka J. Math. **9**, (1972), 217–223.

_______ [75]: On perfect rings and the exchange property, Osaka J. Math. **12**, (1975), 483–491.

Harada, M. and Oshiro, K. [81]: On extending property of direct sums of uniform modules, Osaka J. Math. **18**, (1981), 767–785.

Hausen, J. [79]: Groups whose normal subgroups have minimal supplements, Arch. Math. **32**, (1979), 213–222.

_______ [82]: Supplemented modules over Dedekind domains, Pacific J. Math. **100**, (1982), 387–402.

Hausen, J. and Johnson, J. A. [82]: On supplements in modules, Comment. Math. Univ. St. Paul. **31**, (1982), 29–31.

_______ [83]: A characterization of two classes of Dedekind domains by properties of their modules, Publ. Math. **30**, (1983), 53–55.

_______ [83b]: A new characterization of perfect and semiperfect rings, Bull. Cal. Math. Soc. **75**, (1983), 57–58.

Herrmann, P. [84]: Self–projective modules over valuation rings, Arch. Math. **43**, (1984), 332–339.

_______ [84b]: Projective properties of modules, Algebra Berichte **47**, (1984), Universität München.

Hein, J. [79]: Almost artinian modules, Math. Scand. **45**, (1979), 198–204.

Hermandez, J. L. G. and Pardo, J. L. G. [87]: On endomorphism rings of quasiprojective modules, Math. Z. **196** (1987), 87–108.

Hermandez, J. L. G., Pardo, J. L. G. and Hernandez, M. J. [86]: Semiperfect modules relative to a torsion theory, J. Pure Applied Math. **43**, (1986), 145–172.

Hill, D. A. [73]: Semi–perfect q–rings, Math. Ann. **200**, (1973), 113–121.

_______ [83]: Quasi–projective modules over hereditary noetherian prime rings, Osaka J. Math. **20**, (1983), 767–777.

Holland, S. S. [70]: The current interest in orthomodular lattices, in Trends in Lattice Theory, Van Nostrand, 1970.

Ikeyama, T. [81]: Four–fold torison theories and rings of fractions, Comm. Algebra **9**, (1981), 1027–1037.

Inoue, T. [83]: Sum of hollow modules, Osaka J. Math. **20**, (1983), 331–336.

Ishii, T. [75]: On locally direct summands of modules, Osaka J. Math. **12**, (1975), 473–482.

Ivanov, G. [72]: Non–local rings whose ideals are all quasi–injective, Bull. Austral. Math. Soc. **6**, (1972), 45–52.

Ivanov, A. V. [78]: A problem on abelian groups, Math. USSR Sbornik **34**, (1978), 461–474.

Iwamura, T. [44]: On continuous geometries I, Japan. J. Math. **19**, (1944), 57–71.

Jain, S. K. [76]: Ring theory, Proc. Ohio Univ. Conference, May 1976, Lecture Notes in Pure & App. Math. **25**, Marcel Dekker, 1977.

Jain, S. K., Lopez–Permouth, S. R. and Rizvi, S. T. [P1]: Continuous rings with ACC on essentials are artinian.

_______ [P2]: A characterization of uniserial rings via continuous modules.

Jain, S. K., Mohamed, S. and Singh, S. [69]: Rings in which every right ideal is quasi–injective, Pacific J. Math. **31**, (1969), 73–79.

Jain, S. K. and Mohamed, S. [78]: Rings whose cyclic modules are continuous, J. Indian Math. Soc. **42**, (1978), 197–202.

Jain, S. K. and Müller B. J. [81]: Semiperfect modules whose proper cyclic modules are continuous, Arch. Math. **37**, (1981), 140–143.

Jain, S. K. and Saleh, H. H. [87a]: Rings whose (proper) cyclic modules have cyclic π–injective hulls, Arch. Math. **48**, (1987), 109–115.

—— [87b]: Rings with finitely generated injective (quasi–injective) hulls of cyclic modules, Comm. Algebra **15**, (1987), 1679–1687.

Jain, S. K. and Singh, S. [75]: Rings with quasi–projective left ideals, Pacific J. Math. **60**, (1975), 169–181.

Jain, S. K., Singh, S. and Symonds, G. [76]: Rings whose proper cyclic modules are quasi–injective, Pacific J. Math. **67**, (1976), 461–472.

Jans, J. P. [59]: Projective injective modules, Pacific J. Math. **9**, (1959), 1103–1108.

Jansen, W. G. [78]: Fsp rings and modules, and local modules, Comm. Algebra **6**, (1978), 617–637.

Jeremy, L. [71]: Sur les modules et anneaux quasi–continus, C. R. Acad. Sci. Paris **273**, (1971), 80–83.

_______ [74]: Modules et anneaux quasi–continus, Canad. Math. Bull. **17**, (1974), 217–228.

Johnson, R. E. and Wong, E. T. [61]: Quasi–injective modules and irreducible rings, J. London Math. Soc. **36**, (1961), 260–268.

Kalmbach, G. [83]: Orthomodular lattices, Academic Press , 1983.

Kamal, M. A. [86]: Modules in which complements are summands, Ph.D. thesis, McMaser University, (1986).

Kamal, M. A. and Müller, B. J. [88a]: Extending modules over commutative domains, Osaka J. Math. **25**, (1988), 531–538.

——— [88b]: The structure of extending modules over noetherian rings, Osaka J. Math. **25**, (1988), 539–551.

_______ [88c]: Torsion free extending modules, Osaka J. Math. **25**, (1988).

Kaplansky, I. [51]: Projections in Banach algebras, Ann. Math. **53**, (1951), 235–249.

——— [52]: Modules over Dedekind rings and valuation rings, Trans. Amer. Math. Soc. **72**, (1952), 327–340.

_______ [55]: Any orthocomplemented complete modular lattice is a continuous geometry, Ann. Math. **61**, (1955), 524–541.

_______ [58]: Projective modules, Ann. Math. **68**, (1958), 372–377.

_______ [74]: Commutative Rings, University of Chicago Press, 1974.

Kasch, F. [79]: A decomposition theorem for strongly supplemented and d–continuous modules, McMaster Univ. Math. Reports **105**, (1979), 1–10.

Kasch, F. and Mares, E. [66]: Eine Kennzeichnung semi–perfekter Moduln, Nagoya Math. J. **27**, (1966), 525–529.

Ketkar, R. D. and Vanaja, N. [81a]: A note on FR–perfect modules, Pacific J. Math. **96**, (1981), 141–152.

_______ [81b]: R–projective modules over a semiperfect ring, Canad. Math. Bull. **24**, (1981), 365–367.

Koehler, A. [70a]: Quasi–projective covers and direct sums, Proc. Amer. Math. Soc. **24**, (1970), 655–658.

_______ [70b]: Rings for which every cyclic module is quasi–projective, Math. Ann. **189**, (1970), 311–316.

_______ [71]: Quasi–projective and quasi–injective modules, Pacific J. Math **3**, (1971), 713–820.

_______ [74]: Rings with quasi–injective cyclic modules, Quart. J. Math. Oxford **25**, (1974), 51–55.

Leonard, W. W. [66]: Small modules, Proc. Amer. Math. Soc. **17**, (1966), 527–531.

Li, M. S. and Zelmanowitz, J. M. [P]: On the generalizations of injectivity.

Mares, E. A. [63]: Semiperfect modules, Math Z. **82**, (1963), 347–360.

Matlis, E. [58]: Injective modules over noetherian rings, Pacific J. Math. **8**, (1958), 511–528.

_______ [73]: 1–dimensional Cohen–Macaulay Rings, Lecture Notes in Mathematics **327**, Springer Verlag, 1973.

Michler, G. O. and Villamayor, O. E. [73]: On rings whose simple modules are injective, J. Algebra **25**, (1973), 185–201.

Ming, R. Yue Chi [85]: On von Neumann regular rings, XIII, Ann. Univ. Ferrara Sc. Mat. **31**, (1985), 49–61.

Miyashita, Y. [65]: On quasi–injective modules, J. Fac. Sci. Hokkaido Univ. **18**, (1965), 158–187.

_______ [66]: Quasi–projective modules, perfect modules and a theorem for modular lattices, J. Fac. Sci. Hokkaido Univ. **19**, (1966), 86–110.

Mohamed, S. [70a]: q–rings with chain conditions, J. London Math. Soc. **2**, (1970), 455–460.

_______ [70b]: Semilocal q–rings, Indian J. Pure and App. Math. **1**, (1970), 419–424.

_______ [70c]: Rings whose homomorphic images are q–rings, Pacific J. Math **35**, (1970), 727–735.

_______ [75]: On PCI–rings, J. Univ. Kuwait (Sci). **2**, (1975), 21–23.

_______ [82]: Rings with dual continuous right ideals, J. Austral. Math. Soc. **32**, (1982), 287–294.

Mohamed, S. and Abdul–Karim, F. H. [84]: Semi–dual continuous abelian groups, J. Univ. Kuwait (Sci.) **11**, (1984), 23–27.

Mohamed, S. and Bouhy, T. [77]: Continuous modules, Arabian J. Sci. Eng. **2**, (1977), 107–122.

Mohamed, S. and Müller, B. J. [79]: Decomposition of dual continuous modules, Lecture Notes in Math. **700**, Springer–Verlag, 1979, 87–94.

_______ [81]: Direct sums of dual continuous modules, Math. Z. **178**, (1981), 225–232.

_______ [88a]: Dual continuous modules over commutative noetherian rings, Comm. Algebra **16**, (1988), 1191–1207.

_______ [88b]: Continuous modules have the exchange property, Proc. Perth Conf. Abelian Groups, Contemporary Math. (1988).

Mohamed, S., Müller, B. J. and Singh, S. [85]: Quasi–dual continuous modules, J. Austral. Math. Soc. **39**, (1985), 287–299.

Mohamed, S. and Singh, S. [77]: Generalizations of decomposition theorems known over perfect rings, J. Austral. Math. Soc. **24**, (1977), 496–510.

Mohammed, A. and Sandomierski, F. L. [P]: Complements in projective modules.

Monk, G. S. [72]: A characterization of exchange rings, Proc. Amer. Math. Soc. **35**, (1972), 349–353.

Mucke, C. [88]: Zerlegungseigenschaften von stetigen und quasi–stetigen Moduln, Algebra Berichte **57**, (1988), Universitat München.

Müller, B. J. [68]: Dominant dimension of semi–primary rings, J. Reine Angew. Math. **232**, (1968), 173–179.

_______ [70]: On semiperfect rings, Illinois J. Math. **14**, (1970), 464–467.

_______ [81]: Continuous geometries, continuous regular rings, and continuous modules, Proc. Conference Algebra & Geometry, Kuwait (1981), 49–52.

Müller, B. J. and Rizvi, S. T. [82a]: On the decomposition of continuous modules, Canad. Math. Bull. **25**, (1982), 296–301.

_______ [82b]: On the existence of continuous hulls, Comm. Algebra **10**, (1982), 1819–1838.

_______ [83]: On injective and quasi–continuous modules, J. Pure Applied Algebra **28**, (1983), 197–210.

_______ [84]: Direct sums of indecomposable modules, Osaka J. Math. **21**, (1984), 365–374.

Murray, F. J. and von Neumann, J. [36]: On rings of operators, Ann. Math. **37**, (1936), 116–229.

Nakahara, S. [83]: On a generalization of semiperfect modules, Osaka J. Math. **20**, (1983), 43–50.

Nicholson, W. K. [75]: On semiperfect modules, Canad. J. Math. **18**, (1975), 77–80.

_______ [76]: Semiregular modules and rings, Canad. J. Math. **28**, (1976), 1105–1120.

_______ [77]: Lifting idempotents and exchange rings, Trans. Amer. Math. Soc. **229**, (1977), 269–278.

Oberst, U. und Schneider, H. J. [71]: Die Struktur von projektiven Moduln, Inv. Math. **13**, (1971), 295–304.

Okado, M. [84]: On the decomposition of extending modules, Math. Japonica **29**, (1984), 939–941.

Okado, M. and Oshiro, K. [84]: Remarks on the lifting property of simple modules, Osaka J. Mat. **21**, (1984), 375–385.

Oshiro, K. [80]: An example of a ring whose projective modules have the exchange property, Osaka J. Math. **17**, (1980), 415–420.

_______ [83a]: Semiperfect modules and quasi–semiperfect modules, Osaka J. Math. **20**, (1983), 337–372.

_______ [83b]: Continuous modules and quasi–continuous modules, Osaka J. Math. **20**, (1983), 681–694.

_______ [83c]: Projective modules over Von Neumann regular rings have the finite exchange property, Osaka J. Math. **20**, (1983), 695–699.

_______ [84a]: Lifting modules, extending modules and their applications to QF–rings, Hokkaido Math. J. **13**, (1984), 310–338.

_______ [84b]: Lifting modules, extending modules and their applications to generalized uniserial rings, Hokkaido Math. J. **13**, (1984), 339–346.

Osofsky, B. L. [64]: Rings all whose finitely generated modules are injective, Pacific J. Math. **14**, (1964), 646–650.

_______ [68a]: Non–cyclic injective modules, Proc. Amer. Math. Soc. **19**, (168), 1383–1384.

_______ [68b]: Non–commutative rings whose cyclic modules have cyclic injective hulls, Pacific J. Math. **25**, (1968), 331–340.

_______ [68c]: Endomorphism rings of quasi–injective modules, Canad. J. Math. **20**, (1968), 895–903.

Papp, Z. [59]: On algebraically closed modules, Pub. Math. Debrecen **6**, (1959), 311–327.

Pyle, E. S. [75]: The regular ring and the maximal ring of quotients of a finite Baer *–ring, Trans. Amer. Math. Soc. **203**, (1975), 201–213.

Rangaswamy, K. M. [77]: Modules with finite spanning dimension, Canad. Math. Bull. **20**, (1977), 255–262.

Rangaswamy, K. M. and Vanaja, N. [72]: Quasi–projectives in abelian and modules categories, Pacific J. Math. **43**, (1972), 221–238.

Rayar, M. [71]: Small and co–small modules, Ph.D. thesis, Indiana Univ., (1971).

_______ [82]: On small and cosmall modules, Acta Math. Acad. Sci. Hungar. **39**, (1982), 389–392.

Reda, F. A. [78]: On continuous and dual continuous modules, M.Sc. thesis, Univ. Kuwait, (1978).

Rizvi, S. T. [80]: Contributions to the theory of continuous modules, Ph.D. thesis, McMaster University, (1980).

_______ [88]: Commutative rings for which every continuous module is quasi–injective, Arch. Math. **50**, (1988), 435–442.

Robert, E. de [69]: Projectifs et injectifs relatifs, C. R. Acad. Sci. Paris **286**, (1969), A 361–364.

Roos, J. E. [68]: Sur l'anneau maximal de fractions des AW^*–algebras et des anneaux de Baer, C. R. Acad. Sci. Paris **266**, (1968), A 120–133.

Rudlof, P. [89]: Komplementierte Moduln über Noetherschen Ringen, Ph.D. thesis, Universität München, (1989).

Sakhaev, I. I. [85]: Projectivity of finitely generated flat modules over semilocal rings, Math. Notes **37**, (1985), 85–90.

Sandomierski, F. L. [64]: Relative injectivity and projectivity, Ph.D. thesis, Penn. State. Univ., (1964).

—— [69]: On semiperfect and perfect rings, Proc. Amer. Math. Soc., **21**, (1969), 205–207.

Sarath, B. and Varadarajan, K. [74]: Injectivity of direct sums, Comm. Algebra **1**, (1974), 517–530.

_______ [79]: Dual Goldie dimension II, Comm. Algebra **7**, (1979), 1885–1899.

Satyanarayana, B. [85]: On modules with finite spanning dimension, Proc. Japan. Acad. **61**, (1985), 23–25.

Sharpe, W. D. and Vamos, P. [72]: Injective Modules, Cambridge Univ. Press, 1972.

Singh, S. [67]: On pseudo–injective modules and self–pseudo–injective rings, J. Math. Sci. India **2**, (1967), 23–31.

_______ [80]: Dual continuous modules over Dedekind domains, J. Univ. Kuwait (Sci.) **7**, (1980), 1–9.

_______ [84]: Semi–dual continuous modules over Dedekind domains, J. Univ. Kuwait (Sci) **11**, (1984), 33–39.

Singh, S. and Mehran, H. A. [P]: A note on weak q–rings.

Stenström, B. [75]: Rings of Quotients, Springer Verlag, 1975.

Stock, J. [86]: On rings whose projective modules have the exchange property, J. Algebra **103**, (1986), 437–453.

Suzuki, Y. [68]: On automorphisms of an injective module, Proc. Japan. Acad. **44**, (1968), 120–124.

Szeto, G. [77]: The structure of semiperfect rings, Comm. Algebra **5**, (1977), 219–229.

Swan, R. G. [62]: Vector bundles and projective modules, Trans. Amer. Math. Soc. **105**, (1962), 264–277.

Takeuchi, T. [76]: On confinite dimensional modules, Hokkaido Math. J. **5**, (1973), 1–43.

Utumi, Y. [59]: On a theorem on modular lattices, Proc. Japan. Acad. **35**, (1959), 16–21.

_______ [60]: On continuous regular rings and semisimple self–injective rings, Canad. J. Math. **12**, (1960), 597–605.

_______ [61]: On continuous regular rings, Canad. Math. Bull. **4**, (1961), 63–69.

_______ [65]: On continuous rings and self–injective rings, Trans. Amer. Math. Soc. **118**, (1965), 158–173.

_______ [66]: On the continuity and self injectivity of a complete regular ring, Canad. J. Math. **18**, (1966) 404–412.

_______ [67]: Self–injective rings, J. Algebra **6**, (1967), 56–64.

Varadarajan, K. [79a]: Modules with supplements, Pacific J. Math. **82**, (1979), 559–564.

_______ [79b]: Dual Goldie dimension, Comm. Algebra **7**, (1979), 565–610.

_______ [80]: Study of certain pre–radicals, Comm. Algebra **8**, (1980), 185–209.

Von Neumann, J. [32]: Mathematische Grundlagen der Quantenmechanik, Springer Verlag, 1932.

_______ [36a]: Continuous Geometry, Proc. Nat. Acad. Sci. **22**, (1936), 92–100.

_______ [36b]: Examples of continuous geometries, Proc. Nat. Acad. Sci. **22**, (1936), 101–108.

_______ [36c]: On regular rings, Proc. Nat. Acad. Sci. **22**, (1936), 707–713.

_______ [60]: Continuous Geometries, Princeton Univ. Press, 1960.

Wani, P. R. [86]: Study of the S–module Hom $({}_RM, {}_RN)$ where $S = \mathrm{End}_R M$, Ph.D. thesis, Univ. Calgary, (1986).

Ware, R. [71]: Endomorphism rings of projective modules, Trans. Amer. Math. Soc. **155**, (1971) 233–256.

Ware, R. and Zelmanowitz, Z. [70]: The Jacobson Radical of the endomorphism ring of a projective module, Proc. Amer. Math. Soc. **26**, (1970), 15–20.

Warfield, R. B. [69a]: A Krull–Schmidt theorem for infinite sums of modules, Proc. Amer. Math. Soc. **22**, (1969), 460–465.

_______ [69b]: Decompositions of injective modules, Pacific J. Math. **31**, (1969), 263–276.

_______ [72]: Exchange rings and decompositions of modules, Math. Ann. **199**, (1972), 31–36.

Wisbauer, R. [80]: F–semiperfekte und perfekte Moduln in σ[M], Math. Z. **173**, (1980), 229–234.

Wong, E. T. and Johnson, R. E. [59]: Self–injective rings, Canad. Math. Bull. **2**, (1959), 167–173.

Wu, L. E. T. and Jans, J. P. [67]: On quasi–projectives, Illinois J. Math. **11**, (1967), 439–448.

Yamagata, K. [74a]: The exchange property and direct sums of indecomposable injective modules, Pacific J. Math. **55**, (1974), 301–317.

_______ [74b]: On projective modules with the exchange property, Sci. Rep. Tokyo Kyoiku Daigaku Sec. **A**, (1974), 149–158.

Zariski, O. and Samuel P. [60]: Commutative Algebra II, Van Nostrand, 1960.

Zimmermann–Huisgen, B. and Zimmermann, W. [84]: Classes of modules with the exchange property, J. Algebra **88**, (1984), 416–434.

Zöllner, A. [86]: On modules that complement direct summands, Osaka J. Math. **23**, (1986), 457–459.

Zöschinger, H. [74a]: Komplementierte Moduln über Dedekindringen, J. Algebra **29**, (1974), 42–56.

_______ [74b]: Komplemente als direkte Summanden, Arch. Math. **25**, (1974), 241–253.

_______ [80]: Koatomare Moduln, Math. Z. **170**, (1980), 221–232.

_______ [81]: Projektive Moduln mit endlich erzeugten Radikalfaktormoduln, Math. Ann. **255**, (1981), 199–206.

_______ [82a]: Komplemente als direkte Summanden II, Arch. Math. **38**, (1982), 324–334.

_______ [82b]: Gelfandringe und koabgeschlossene Untermoduln, Bayer. Akad. Wiss., (1982), 43–70.

_______ [83]: Linear–kompakte Moduln über noetherschen Ringen, Arch. Math. **41**, (1983), 121–130.

_______ [86]: Komplemente als direkte Summanden III, Arch. Math. **46**, (1986), 125–132.

NOTATION

ring	ring with identity
module	unitary right module
summand	direct summand
lsTn	locally–semi–transfinitely–nilpotent
$\mathbb{N}$	$\{1,2,3,...\}$
$\mathbb{Z}$	ring of integers
$\mathbb{Q}$	field of rational numbers
$\mathbb{R}$	field of real numbers
$\mathbb{C}$	field of complete numbers
$\hat{\mathbb{Z}}_p$	ring of p–adic integers
C_{p^n}	cyclic group of order p^n
C_{p^∞}	Prüfer group
$\subseteq$	set inclusion
$\leq$	submodule
$<$	proper submodule
$\leq^e$	essential submodule
$<<$	small submodule
$\subset^\oplus$	summand
$\longrightarrow$	homomorphism
$\rightarrowtail$	monomorphism, embedding
$\twoheadrightarrow$	epimorphism
$\rightarrowtail\!\!\!\!\twoheadrightarrow$	isomorphism
$S = \text{End } M$	ring of endomorphisms of M
Δ	$\{f \in S : fM \leq^e M\}$
∇	$\{f \in S : \ker f << M\}$
RadM	Jacobson radical of M
J(R) or J	Jacobson radical of a ring R
E(M)	injective hull of M
X^o	annihilator of X
$\prod_{i\in I} M_i$	direct product
$\bigoplus_{i\in I} M_i$	direct sum
M(K)	$\bigoplus_{i\in K} M_i$, for $K \subseteq I$

$M(I\text{–}i)$	$M(I - \{i\})$, for $i \in I$
$X^{(n)}$	direct sum of n copies of X, for a cardinal n
$X^{(\mathbb{N})}$	$X^{(n)}$, for $n = \mathbb{N}$
$\mathscr{X}^{\perp}$	the orthogonal class of $\mathscr{X}$
$\Rightarrow$	implies
$\Leftrightarrow$	equivalent
$\square$	end of proof

INDEX